KNOWLEDGE AND REALITY

THE WESTERN ONTARIO SERIES IN PHILOSOPHY OF SCIENCE

A SERIES OF BOOKS
IN PHILOSOPHY OF SCIENCE, METHODOLOGY, EPISTEMOLOGY,
LOGIC, HISTORY OF SCIENCE, AND RELATED FIELDS

VOLUME 59

PAOLO PARRINI
Department of Philosophy,
University of Florence

KNOWLEDGE AND REALITY

An Essay in Positive Philosophy

translated by

Paolo Baracchi
Department of Philosophy,
University of Melbourne

KLUWER ACADEMIC PUBLISHERS
DORDRECHT / BOSTON / LONDON

A C.I.P. Catalogue record for this book is available from the Library of Congress.

ISBN 978-90-481-4990-2

Published by Kluwer Academic Publishers,
P.O. Box 17, 3300 AA Dordrecht, The Netherlands.

Sold and distributed in North, Central and South America
by Kluwer Academic Publishers,
101 Philip Drive, Norwell, MA 02061, U.S.A.

In all other countries, sold and distributed
by Kluwer Academic Publishers,
P.O. Box 322, 3300 AH Dordrecht, The Netherlands.

Printed on acid-free paper

To Simonetta

Mut

Will kein Gott auf Erden sein,
sind wir selber Götter!

[*Courage*

If there is no God on earth,
then we are Gods ourselves!]
(F. Schubert, W. Müller, 'Winterreise')

TABLE OF CONTENTS

FOREWORD TO THE ENGLISH EDITION

This book – a somewhat modified version of my *Conoscenza e Realtà* (1995) – is the result of many years of reflection and research in philosophy of science, epistemology, and in the history of scientific and philosophical thought.

The main historical thrust of my work is a specific conception of logical empiricism and its relations with Kantian and conventionalist philosophies. The revision of the traditional interpretation of the philosophy of the Vienna and Berlin Circles is today generally accepted, and it has solid roots in many European countries as well as the United States. In contrast, when I first outlined a new reconstruction of the German and, more generally, 'continental', origins of the neopositivist movement at the beginning of the seventies, certain of my theses were considered – at least in Italy – highly provocative.[1]

The close connection with my historical research is one of the two main reasons for the somewhat unusual form of my book, especially when compared to the standards of Anglo-American works. Throughout, I have developed my theoretical proposal – that of a positive, although not strictly positivist philosophy – by means of a dialectical comparison with relevant positions, present and past. Taking on board Rorty's idea of philosophy as the "conversation of mankind",[2] I have tried to prove that the positive philosopher can join in the conversation as well.

The second reason that induced me to give the book its peculiar form is connected to my conception of philosophy and of the role of philosophical analysis.[3] The essential point is my conviction – supported precisely by the analysis of the problem of knowledge – that it is not possible to develop a comprehensive philosophical discourse whilst limiting ourselves to the mere analysis of language, of problems, or of concepts. As I maintain in the final pages of Chapter VII, with reference to Herbart, as well as the aspect of analysis, there must be an aspect of 'synthetic-rational reworking' (*Bearbeitung*), of 'conceptual engineering'. The 'synthetic' aspect consists in the proposal of a theoretical construction which is capable of partly modifying the initial concepts when this is necessary. This affects the structure and presentation of philosophical work, making the latter very close to the construction of scientific theories.[4]

---

[1] See Parrini 1980b; 1983, 1987; 1994b; 1993; 1994c; 199+a. I use the expression 'continental philosophy' (e.g., in Parrini 1993) in the sense in which it is used in Solomon (1988): I stress the affinities which logical empiricism has with the 'continental' tradition rather than its affinities with the British empiricist tradition, which are in the forefront, e g., in Ayer (1962).

[2] Rorty 1980· 389ff.

[3] See Parrini 199+b, sec 3 / 2

[4] As is well known, there are as many ways of practicing philosophical analysis as there are major exponents of analytic philosophy. For example, Russell does not proceed

The stability of a philosophical construction will depend not only upon the solidity of the blocks, of the pillars and architraves that make it up, but also upon the way in which all these parts are connected. Of course, it will not be possible to argue for every single part of a philosophical building: to do so would mean to embark in a virtually endless enterprise. Accordingly, some of the parts of a philosophical building will have to be taken from the literature on the subject as 'ready made' or 'semi-finished' elements, while others will be argued for in the course of building.

This is what happened in my work too. In some cases (for instance, in the case of epistemic relativism), my concern was to illustrate theses which I believed to be sufficiently consolidated, rather than to argue for them. In other cases – where I was directly engaged in building the theory that I want to formulate – I did exactly the opposite. This is what I have tried to achieve, for example, for those proper architraves of my construction, viz. the connection between scepticism and metaphysical realism, and the thesis of the normative value of the fundamental epistemological notions (truth, objectivity, and rationality).

The main theoretical motivation that guided me in the construction of this building was the conviction that positive philosophy should free itself from semantic conceptions such as the verification principle, and reconnect itself to the traditional formulations of the positivist critique of metaphysics. I embrace Mach's idea that the anti-metaphysical attitude is in the first place an anti-absolutist attitude. The main difference with Mach's position is that I have tried to develop this idea by connecting it not to sensationalist phenomenalism, but rather to a form of non-reductive empiricism, that is, to an empiricism which admits the reality (albeit not in the sense of *metaphysical* realism) of the objects of experience. The essential feature of this empiricism is the rejection of Kant's theory of the synthetic a priori, in accordance with Schlick's and Carnap's tenet: empiricism is fundamentally characterised by the rejection of the synthetic a priori.[1]

My relationship with Kant is complex. It is not mere rejection. On the contrary, my proposal could even be called critical-positive, since it re-elaborates the Kantian and neo-Kantian (especially Cassirer's) conception of knowledge, truth and objectivity. In my view, if we are to progress beyond the traditional opposition between metaphysical realisms and anti-realisms, thus promoting a new positive philosophy,

---

in the same way as the Wittgenstein of the *Philosophical Investigations*, Moore does not proceed in the same way as Quine, and, even within logical empiricism proper, Carnap does not proceed in the same way as Reichenbach. In a letter to Maria Reichenbach dated August 18th 1975 [HR-041-019-015], Salmon defends Reichenbach from the accusation of not having been consistently rigorous, by saying that, contrary to authors such as Carnap and Quine, Reichenbach's way of proceeding and his insight were more like those of a physicist than like those of a mathematician. Personally, I feel closer to Reichenbach's style than to Carnap's

[1] Schlick 1974: 384, 398f; Carnap 1995. 180

the neoempiricist critique of the synthetic a priori must be connected with some aspects of Kant's theory of truth and objectivity.

I am aware of the fact that nowadays the realism / anti-realism problem is construed by many philosophers in terms of logical and semantical contrapositions, where verificationism turns out to be connected with the logical principle of bivalence. I do not think that this is the best way of addressing the issue: I see the problem of realism and anti-realism as primarily connected with the problem of knowledge and with the philosophy of science. In particular, I am keen to propose an early twentieth century formulation of the epistemological problem, which hinged on the controversy between immanent and transcendent conceptions of the object of knowledge.

The immanent perspective is well described by Husserl: "Knowledge must be investigated in itself, in its proper essence, not in relation to a presupposed real world, and not as a fact in this world, but in pure immanence, as absolute givenness: only in this pure immanence and absolute givenness is the totality of the real world (a world which is a represented, judged, known world, and which is only by virtue of this connection the 'real world' for knowledge) 'constituted' (a word, this, which must certainly be understood in a very precise manner)".[1]

According to the intentions of Husserl and his followers, this purely immanent analysis should serve to elaborate a theory of knowledge independent of any presupposition about the reality of the object and of the subject of knowledge. This corresponds to the Marburg reading of the *Critique of Pure Reason*, which Cassirer illustrated by means of Schiller's famous epigram: the Critical approach to the epistemological problem knows nothing, preliminarily, of the soul and nothing of things.[2]

It is now a well-established historical fact that this perspective on knowledge, despite its neo-Kantian and phenomenological origins, was not foreign to neoempiricism. For example, neo-Kantianism influenced both Reichenbach's 1920 book *Relativitätstheorie und Erkenntnis A priori*, and Carnap's position in *Der logische Aufbau der Welt* (1928). Moreover, in this work, Carnap defends the 'neutrality' (in Husserl's sense) of the constructional system with respect to traditional metaphysical options (realism and idealism, phenomenalism and materialism, spiritualism and solipsism), and introduces the distinction between empirical and metaphysical reality – and therefore between empirical and metaphysical realism. The same distinction was taken up by Schlick in 'Positivismus und Realismus' (1932), abandoning the transcendent realism, which he had defended in the *Allgemeine Erkenntnislehre* (1918, 1925) precisely against the *Immanenzgedanken*, that is, the philosophies of immanence.[3]

---

[1] Husserl 1987 5.
[2] Cassirer 1922. 662.
[3] See Parrini 1993, Part II, 'Realismo, strumentalismo, neopositivismo', pp. 377-93, and Parrini, 199+a.

The logical empiricists' views on realism were neither uniform nor static. Reichenbach, modifying what he had maintained in the twenties, objected to Carnap's *Aufbau* that the alleged metaphysical neutrality of the *Konstitutionssystem* was only 'a nice dream' [*ein schöner Traum*], and that it was not possible to do without the *Axiom des Realismus*. On the other hand, Carnap limited himself to noting in shorthand, in the margin of the letter in which Reichenbach made this remark, that "*In* science, however, the axiom of realism cannot be formulated".[1] During the controversy over protocols, Schlick, once again defending a realist point of view, argued against Neurath for the necessity of the idea of the agreement between reality and those peculiar assertions, the *Konstatierungen*, "which record 'facts of immediate observation'".[2] In his turn, Neurath replied that "Schlick's accusation obviously rests on the fact that he speaks of 'the' one reality that can be described only by one of several irreconcilable systems of statements', whereas we stress that this formulation does not occur within scientific language, but that the task is to select one among several possibilities".[3]

However, reference to the verifiability principle was an essential feature of all these debates (with the partial exception of Reichenbach); the cogency of Carnap's and Neurath's criticisms of metaphysical realism was dependent upon their reliance on the verifiability principle. The subsequent crisis of the verifiability principle weakened Carnap's reply to Reichenbach and strengthened Schlick's objection to Neurath. That is, both Reichenbach's axiom of realism and Schlick's idea of agreement with reality can no longer be considered as devoid of meaning. After the crisis of the verification principle, all those who have followed Carnap and Reichenbach, Schlick and Neurath in the tradition of analytic and scientific philosophy are inevitably faced again with the problem of realism. Thus the present epistemological debate brings to the fore once again the contrast between metaphysical realists, instrumentalists and upholders of more or less markedly relativist theses connected with the idea of the plurality of worlds and truths.

In these recent discussions, one epistemological tradition seems to have been forgotten: the tradition that I just mentioned, which tried to coherently combine immanence and transcendence in the cognitive process. Connecting myself to this tradition, I have endeavoured to build a philosophical proposal which reconciles empiricism with a Critical, anti-correspondentist conception of truth and objectivity. I have tried to trace a third way between the Scylla of metaphysical realism and the Charybdis of radical relativism: I wanted to account for the aspects of subjectivity present in the cognitive and, in particular, in the scientific enterprise, without giving up all claims to objective validity. My enter-

---

[1] "*In* der Wissenschaft kann man aber das Realitätsaxiom nicht aussprechen" (I thank W G. Heverly for the transcription). See Parrini 1993, Part II, 'Realismo, strumentalismo, neopositivismo', pp. 380f and n. 8.

[2] Schlick 1979d: 375f; see also 1979e: 407, 409.

[3] Neurath 1983 106

prise should prove that, contrary to what is often assumed, one can refuse to equate being true with being believed to be true without committing oneself to metaphysical correspondentism, i.e., to the thesis that, for a statement to be true, there must be something in reality in itself which makes it true.

The arguments for my 'third way' turn upon two essential pivots: (1) knowledge is unification of experience, not the mirroring of a transcendent ('*ex-manent*') reality; (2) truth and objectivity are empty regulative ideals guiding a search that intentionally aims at achieving them. On this basis, I repropose, in a totally non-scientistic form, the guiding ideals of the *esprit positif*: empiricism, moderate epistemic relativism, anti-metaphysical objectivism. It is my hope that such a 'positive philosophy' will enable us to avoid the unsatisfactory positions of radical relativism, reinstating science as the paradigm of cognitive and explanatory activity. I do not think that we will thereby find ourselves committed to the idea – rightly criticised by postmodern philosophers – of a single overarching grand narrative or of a privileged language capable of saying everything about its referent.

I would like to close with my sincere thanks to all those who, in recent years, have variously encouraged and helped me to make my ideas known to the English-speaking public: Martin Carrier, Michael Friedman, Adolf Grünbaum, John Haldane, Don Howard, Peter Machamer, David Malament, Wes Salmon, Michael Stölzner, Gereon Wolters, and the missed Robert Butts, who warmly accepted the idea of publishing this book in the collection directed by him. A particular thanks to Howard Sankey and Stephen Barker, who read the English version of the book, giving valuable advice. Finally, a warm thanks to Paolo Baracchi for the weighty and delicate work of translation.

Florence, June 1997

Paolo Parrini

TRANSLATOR'S NOTE

My effort has been to achieve the maximum clarity in arguments and directness in expression. To this effect, and keeping in mind the public that this English edition is aimed at, it seemed worthwhile to sacrifice to a certain extent the original, typically Italian, style of Parrini's book, with its constant awareness of the theoretical and historical complexity and interrelatedness of the topics, and its keen perception of conceptual details, facets and nuances. While trying to preserve the full conceptual articulation of the arguments, I have somewhat simplified the original vocabulary and syntax; I have broken down the original paragraphs into shorter units; I have divided each chapter into sections and subsections, to which I have assigned titles; throughout I have signposted the structure of Parrini's arguments by making free use of internal numberings. What is lost in elegance and smoothness of expression is – I hope – gained in perspicuousness of the logical structure of the work.

The Italian edition appeared without references or bibliography: the citations were accompanied only by the name of the author and, sometimes, the title of the work they were taken from. There were no footnotes. For the English edition, it was decided to follow more usual criteria. The author supplied me with 261 endnotes, containing most of the bibliographical references, and further remarks. Each citation was checked against the original, and the bibliography of the volume was put together. I have endeavoured to provide the reader with a scholarly apparatus which is rigorous, easy to consult, and not too cumbersome.

It will be noted that the standard practice of printing the longer quotations as separate paragraphs, indented and often in smaller type, has not been followed. This is an anomaly of the Italian edition which I decided not to change: I thought that typographical uniformity would help preserve the close blend, in some cases amounting almost to a fusion, between the text and the numerous quotations – a salient feature of the original volume.

I also compiled the index of names and subjects.

All my interventions were done with the consent and under the supervision of the author.

While taking full responsibility for the defects and shortcomings of my work, I wish to express my indebtedness to Dr. Howard Sankey, of the Department of History and Philosophy of Science of the University of Melbourne, who meticulously read through a draft of the whole translation, giving me detailed feedback on particular problems, as well as discussing with me a number of general issues. My gratitude goes also to Dr. Stephen Barker, who revised another draft of my translation, and to Ms Siu Chan for her valuable assistance in compiling the bibliography and checking the quotations. Last but not least, I wish to mention the stimulating and friendly atmosphere of Melbourne University's Philosophy Department, where I carried out most of my work: my sincere appreciation to staff and students.

Paolo Baracchi

INTRODUCTION

1. SAILORS AND GODS. REASON, EXPERIENCE, AND PHILOSOPHICAL ANALYSIS

In Plato's *Phaedo*, Simmias, speaking about the immortality of the soul, says: "I think, Socrates, as perhaps you do yourself, that it is either impossible or very difficult to acquire clear knowledge about these matters in this life. And yet he is a weakling who does not test in every way what is said about them and persevere until he is worn out by studying them on every side. For he must do one of two things; either he must learn or discover the truth about these matters, or if that is impossible, he must take whatever human doctrine is best and hardest to disprove and, embarking upon it as upon a raft, sail upon it through life in the midst of dangers, unless he can sail upon some stronger vessel, some divine revelation, and make his voyage more safely and securely."[1]

This book is born from an entirely secular persuasion, though neither adverse nor insensitive to religious feeling. This persuasion is that, in order to cross the sea of life, no better means are available to human beings than those they can build themselves using reason and experience. Adapting an old metaphor, originally Neurath's, later adjusted by Quine and Putnam, we could say that human beings are like sailors who have a certain number of more or less well rigged ships. These must be repaired during the journey, for it is not possible to bring them all together in a shipyard to rebuild them from top to bottom with imperishable and unchangeable materials. During the journey, there are no absolute points of reference; it is an illusion to think that supposed messages of divine origin might convey certain and unquestionable directions about the route to follow. The interpretation of the revealed word and its adaptation to the changing historical circumstances have proved themselves to be such problematical enterprises, that it looks like humanity has never had any real alternative to relying upon the best and least disputable of human arguments. Resort to divine revelation can spare us none of the difficulties with which we have to grapple; we must rely on our experience and rational capacities. Intersubjective inaccessibility and inoperativity seem to be basic traits of the Absolute, in moral life as in cognitive practices.

This aspect of the *positive* spirit has not been affected by all the objections recently raised against *positivist* philosophies. Nevertheless, it has become increasingly necessary to make space for the subjective components of moral and cognitive activity. Be it a matter of judgements of plausibility or of behavioural choices, it seems impossible to exclude all reference to subjective evaluative criteria, which, moreover, seem irreducible to explicit discursive rules. This is one of the major teachings bequeathed to us by probability theorists of various orienta-

[1] Plato 1914: 297 [I, 85, C-D].

tions, such as Keynes and De Finetti, and by upholders of tacit knowledge, such as Polanyi.

My arguments will be interwoven with elements from different philosophical traditions. In the present 'conversational' philosophical climate, this should be a merit; however, some readers will probably see this as yet another manifestation of the mediocre eclecticism of recent Italian philosophy. A considerable part of Italian post-war philosophical production has been accused of eclecticism. We should be wary of this indictment. Many recent historical reconstructions of that philosophy are philosophically so weak that, in comparison, even the most simplistic versions of certain past Italian eclecticism are archetypes of conceptual complexity. As for me, I have tried to set forth sound arguments for a rigorous conceptual elaboration. I have not looked for facile juxtapositions, but for a genuine connection between divergent trends of thought.

As a matter of fact, another reason which prompted me to write this book is the conviction that the great analytic effort spent over the last fifty years in criticising the fundamental theses of logical empiricism has not been followed by a comparable analytic effort in the reconstruction of an alternative conceptual framework. One of the most fascinating characteristics of logical empiricism was its tendency to conjugate general philosophical formulations (such as the verification principle and the linguistic conception of the a priori) with detailed particular analyses (concerning, for example, scientific explanation, inductive confirmation, the structure of theories, etc.). This tendency has progressively waned in later analytic philosophy, leaving an almost exclusive attention to specific questions. Carnap himself regretted this: in the correspondence of his last years, he repeatedly deplores the young analytical philosophers' disinclination for 'foundational' philosophy. The ambition of this book is the analytically constructive search for a general perspective that may work as a unifying framework for specialised analyses in epistemology and in the philosophy of science.

The analytic style, from which in many ways I continue to draw my inspiration, is no longer as popular as it was some decades ago, when it characterised a large part of mainstream philosophical activity. Some contemporary philosophers have accused it of fostering a dangerous inhibition of speculative creativity. Some have even spoken about it as an almost masochistic 'desertifying' tendency. These criticisms seem to me excessive and misdirected. Philosophy has always had a certain theoretical responsibility: only by not giving up the claim of being theory (with all the constraints that this entails), can philosophy distinguish itself from entirely ideological and subjective structures. And since the times of Socrates, whom Schlick once identified as the father of philosophical analysis,[1] the analytic aspect has shown itself to be intrinsically tied to philosophy's claims to theoretical credibility.

Giving up analytic commitment may lead to a pleasant loosening of inhibitions, to an exciting state of theoretical irresponsibility.

---

[1] Schlick 1979f. 496f.

Once on this path, we might even claim to do philosophy whilst abandoning the form of theoretical argument altogether in favour of narrative expression. In some cases, there has been no hesitation to embrace the bizarre thesis of the replacement of theory by narration as an indispensable condition to find the 'courage' to address 'the great problems of metaphysics'. In this way, it will be easier than it was for Freud in his late maturity to unbridle one's speculative and ideological impulses. The danger (largely already come true) will be that of getting results as insipient theoretically as they are narratively boring. Surely, it is possible that neither philosophy nor science will ever develop a creativity and depth comparable to poetry and art. But this should be a further reason to respect the autonomy of poetical and literary creation, shunning the temptation of using them as shortcuts for our theoretical impotence.

2. DWARFS AND GIANTS. PHILOSOPHY AND HISTORY

An essential part of the book deals with the cognitive value of scientific theories, and consequently belongs to the philosophy of science. However, my overall proposal aspires to a wider inclusiveness, concerning epistemology and philosophy in general.

I develop my arguments by means of analytic comparisons with significant positions of the past. Accordingly, I take into consideration the historical dimension of the problems. I chose this method for two reasons.

(1) In the first place, the method of historical comparison brings into a particularly sharp focus the epistemological and ontological significance of some specific problems, whose treatment in the textbooks of philosophy of science is generally purely systematic. I have long been convinced of the necessity of integrating historical research and theoretical elaboration. I am increasingly persuaded of the heuristic value of 'historical-critical' investigation as a means of clarifying the fundamental concepts of a discipline and promoting the solution of its problems. As Mach and Duhem saw, often in philosophy, as in the sciences, to retrace the history of a subject means to make a sort of logical analysis of it. The life of concepts is interwoven with accidental historical circumstances as much as with subtle logical interconnections, and we should try hard not to lose sight of the relations between the two planes.[1]

(2) Secondly, I follow the method of historical comparison because I share Santayana's (and Laudan's) idea that "[t]hose who cannot remember the past are condemned to repeat it".[2] This holds particularly in an age of epigones such as ours. In the present phase of thought, we can consider ourselves, at best, as dwarfs on giants' shoulders. Although

---

[1] See, e.g., Mach 1974: xxiif, 1; Duhem 1962a, p. II, ch VII, sec 6, 'The importance in physics of the historical method' (pp. 268-70).

[2] Santayana 1962: 184.

I am convinced of this, I am unwilling to reduce philosophy to a form of theoretically unspecified historical knowledge. Our aspiration to climb onto the giants' shoulders requires us to take on precise evaluative (and therefore theoretical) responsibilities: this we must do if we want to be able to identify the highest points of our tradition of thought. It would be sad if the thought that we possessed a privileged perspective reconciled us to our theoretical weakness, only to discover later that we had taken the curve of the giant's big toe for that of his shoulder. If our being epigones does not justify a theoretically non-committed form of historicism, it would be equally naïve to believe it possible to define our theoretical position without considering the past.

One of my main aims is to avoid the two extremes.

The first is a purely structural and foundational conception of theoretical justification, which leaves aside the evaluative parameters offered by the historical context.

The other extreme is a conceptually 'invertebrate' historicism (the expression is not mine) which ends up by making not only objectivity, but even intersubjectivity, impossible. In history there is everything and the contrary of everything; therefore, without (historically conditioned) parameters of theoretical evaluation, no justificational use of history is possible. The fact that no moorings are available to us in the form of absolutely founded norms of judgement does not entail that we are to float adrift in the currents of history, renouncing all claims to conceptually structured judgements. I think that the now manifest emptiness of many recent discussions on modernism and postmodernism makes my point.

3. TOWERS AND LADDERS. PHILOSOPHY, ANALYSIS, AND SCIENCE

The question of objectivity concerns also the validity that philosophical discourse can aspire to.

Neurath used to say that, while logical analysis unites the different areas of investigation, traditional philosophy leads to the construction of towers from which philosophers, like medieval squires in their fortified houses, fight amongst each other and raise their fortifications, thus making understanding and cooperation increasingly difficult.[1]

In the years during which Neurath upheld this view, there was strong hope of achieving a 'scientific philosophy'. Its aim was to realise within the philosophical world a relative uniformity of views, similar to that which characterises the scientific community. It would be a mistake to deny any success to the programme of making philosophy increasingly scientific. This programme was pursued especially by the logical positivists, and it is largely to their merit that our understanding of the structure of philosophical problems is today so much enhanced.

---

[1] See Neurath 1935 53.

Nevertheless, we must recognise that a marked consensus was not achieved even among the neopositivists themselves, let alone among philosophers in general.

An important aspect of my proposal is the acknowledgment that the nature of philosophical problems is such that it is impossible to construct a conceptual system that leaves no room for elements of option and arbitrariness. The danger of barricading ourselves in towered fortifications cannot be wholly excluded. However, I believe we can, and must, set ourselves the task of reducing as much as possible the room for theoretical options which cannot be intersubjectively examined according to reason and experience (in the sense which I will specify in the last chapter). I think we can still share a large part of our philosophical tradition's aspiration to construct a building as little turreted as possible, and providing some guarantee of solidity.

If anything, the problem is why address highly controversial and doubtful philosophical problems on which, as Clifford and Mach wittily point out, "it is peculiarly difficult to make out precisely what another man means, and even what one means one's self".[1] It would be the highest aspiration of positive philosophy to avoid the grounds of traditional philosophical disputes, which risk falling into nonsense, and keep to logical and empirical analyses that show that it is not meaningfully possible to go beyond their results. This aim is clearly expressed in the well-known proposition 6.53 of the *Tractatus*: "The correct method in philosophy would really be the following: to say nothing except what can be said, i.e. propositions of natural science – i.e. something that has nothing to do with philosophy – and then, whenever someone else wanted to say something metaphysical, to demonstrate to him that he had failed to give a meaning to certain signs in his propositions".[2] The positive philosopher's ideal would be to attain this goal without even using *ladders* to be thrown away after having been climbed over.

Unfortunately, the failure of the absolute criterion of meaningfulness proves that such an enterprise is unrealisable. Consequently, positive philosophers are unhappily forced to go beyond the scope of the sheer analysis of concepts and facts, and become, against their own inclinations, philosophers in the full sense. They will strive to preserve as much as possible of the positive spirit, shunning the traditional forms of positivism which have not withstood the test of philosophical analysis.

Of course, it would be possible to avoid such slippery ground by simply choosing not to ascend to the plane of philosophical generality, and restricting one's interests to specific historical and theoretical analyses. One could rest content with the moderate amount of more or less incidental generalisation that such analyses require. Personally, I feel as much sympathy for this kind of work as I feel instinctive disinclination for the discourse of a general kind. To think that this is at odds with my profession would show unfamiliarity with the analytic specificity re-

[1] Clifford 1879: 88, cited in Mach 1959: xxxix.
[2] Wittgenstein 1994. 73f.

quired by a large part of philosophical investigation. It is possible to practise the "honest work" of the philosopher which Giulio Preti talked about[1] while sharing Austin's opinion that philosophers indulge too easily in unwarranted generalisations. However, in the end I convinced myself that I had to overcome my aversion.

It is clear that the question I am addressing concerns the reasons for philosophical discourse, in the full range of its implications. One of the most frequently given answers is Popper's. Since all human beings are involved in philosophy, because in different ways they all have an attitude towards life and death, we must have the courage to oppose "the minute philosophers with their minute problems" (i.e., the analytic philosophers of common and of scientific language). This is necessary in order to engage in the principal task of philosophical research, that is, "to speculate critically about the universe and about our place in the universe, including our powers of knowing and our powers for good and evil". For Popper, philosophy "lives in the great problems which beset mankind and which demand new and bold [...] thought"; it dies in scholasticism where "all the great ideas are buried in a flood of words".[2]

These expressions have always seemed to me serious misunderstandings of the real issue raised by analytic philosophers (neopositivists and non-neopositivists). Rather than deny the great questions, they probably sought to relegate them to a domain which is not properly cognitive. Moreover, Popper underestimates the possibility that there may be important connections between 'minute' and great questions. For these reasons, I have always preferred to Popper's rather pompous words the ironically sober remarks with which the physicist Campbell tried to persuade his scientist colleagues of the scientific importance of 'metaphysical' investigation. After having pointed out that "we are all metaphysicians, physicists included", he declared that "[t]he world is not divided into those who do and those who do not hold metaphysical doctrines, but rather into those who hold them for some reason and those who hold them for none".[3]

A few decades ago, I would have quoted this sentence with a certain confidence, to clarify and defend the aim of my book. Now, I am less confident, because among philosophers themselves a seemingly increasing number believe that it is not very important to give reasons for what we uphold, implicitly or explicitly, when dealing with philosophical problems.

Worse still, as an advocate of a positive (even if not positivist) philosophy, I am not in the best tactical position to oppose philosophical defeatism. The strongest reasons which, as a positive philosopher, I am willing to admit are logical-analytic arguments, empirical proofs or a mixture of the two. Yet, I must resign myself to the idea that my positive

---

[1] Preti 1976a: 478.
[2] Popper 1986: 210f.
[3] Campbell 1957: 12.

philosophy goes beyond what can be argued for in this way – even if this does not entail that I have nothing to offer in its defence. All this will be spelt out in the last chapter. The essential point, however, is that our 'philosophical towers' cannot be built using only strictly logical and empirical materials. We need also an overall evaluation which leaves room for margins of subjective arbitrariness and preference.

For these reasons, in explaining why I have decided to outline a positive philosophy, I must mention a motivation which is to a certain extent personal. The book actually came into being as a reflection upon the principles informing the kind of philosophical analysis which I have been practicing over the past twenty years. One of the book's main goals is the elaboration of a general theoretical perspective, in the sense of a conceptual frame of reference. Looking at many specific questions I have (or would like to have) addressed, I realised that I was unable to lead particular analyses to the desired conclusions because I did not possess the necessary general framework. I came up against this problem for the first time more than ten years ago, when I was invited by some friends to give a talk on Kripke's theory of reference. I realised the impossibility of disputing the theory's possible metaphysical import, while limiting myself to an internal analysis which made no reference to an alternative philosophical perspective.

As the saying goes, *tout se tient*. But then the problem – and what a problem! – becomes that of understanding which are the joints that hold the whole together. This seemed an exquisitely philosophical task, in the sense of the philosophical component Mach vindicated for science. It is true that Mach proclaimed himself contrary to the intrusion of the philosophical method in science, and did not want to be considered a philosopher; but he recognised the presence of a philosophical element in science, in so far as scientists aim at "bringing their single conceptions into agreement with the general conception or ideal of an entire province of phenomena" constantly considering "the whole in their treatment of parts".[1] As the neopositivist Frank suggested a few years later, referring to Aristotle and Comte, there is space for philosophical investigation wherever there is need of a form of conceptual integration.[2] And, if a scientist can perhaps 'acquiesce' in Einstein's 'methodological opportunism',[3] philosophical 'acquiescence' seems to require something conceptually more structured.

To use a speculative terminology, for me the big traditional problem of reconciling the particular with the universal still holds. My goal is a reconciliation that (a) does not dissolve the particular in the universal, and (b) traces the universal back to the abstract epistemic conditions which, in an historically given context, grant the particular a relatively justifiable conceptual comprehension of other particularities, including itself. One of the most pleasant surprises in the course of the

[1] Mach 1974: 39.
[2] Frank 1975 37.
[3] See Einstein 1951: 684.

work was to see a personal experience confirmed and clarified in the form of a theoretical conclusion: positive philosophy feeds on the recognition of a quasi-circular interdependence between general theorisation and particular analysis. The partial failure of logical empiricism proves that the positive spirit cannot be confined in abstract general formulae (such as the verification principle, or the distinction between analytic and synthetic statements) with no contextual specifications. Contextual specifications are needed in order to supply concrete contents relative, for example, to certain historically determined forms of scientific knowledge, to particular disciplines and theories of the exact and empirical sciences, to specific controversies concerning the scientificity of an hypothesis, etc.

I should like it to conclude by making it clear that the present outline of a positive philosophy is not an attempt to defend scientific knowledge against the recent attacks from various areas of philosophy and epistemology. I agree with Campbell in thinking that the attitude of the philosopher who wants to defend science (an attitude not wholly foreign to logical empiricism) can be more offensive to science than the attitude of one who attacks it. To be more precise, I believe that philosophers (even great ones) who attack or defend science in an exclusively philosophical perspective set themselves a task that is more preposterous than offensive. They forget that science – which, of course, is a different thing from what scientists say when they make philosophical extrapolations – has asserted itself, from Descartes onwards, by developing a procedure which sets aside abstractly philosophical questions. Campbell's observation is still valid: both defenders and detractors of scientific knowledge "seldom took the trouble to understand what they attacked" or defended.[1]

I really do not think that we can talk of a crisis of science brought about by philosophical criticism. Rather, I am persuaded of the contrary. The development of science has raised such serious difficulties for philosophy, that the survival of the latter has become problematic – at least for the parts of it dealing with epistemological and ontological problems. If, in the centuries that divide us from Galileo, something has met with increasing difficulties, this is philosophy, not science. Be this as it may, crisis or non-crisis of philosophy, death or non-death of philosophy, it is the new reality constituted by the growing scientific knowledge that philosophical speculation has had to acknowledge, in order not to lose all connections with the cultural universe we have been inhabiting for the last few centuries.

[1] See Campbell 1957: 9.

Chapter I

POSITIVISMS OLD AND NEW

1. CRISIS OF THE POSITIVE SPIRIT OR CRISIS OF NEOPOSITIVISM?

1.1 THE STANDARD CONCEPTION AND THE *WISSENSCHAFTLICHE WELTAUFFASSUNG*

The last forty years of research in the philosophy of science have been characterised by the crisis of the standard conception of scientific theories. This model was developed by the logical positivists, especially Carnap and Hempel, over a long period of time: from the formation of the Vienna Circle in the late twenties, up to the end of the process of liberalisation of empiricism in the late sixties (roughly when the new philosophy of science was asserting itself).

Technically, the standard conception was the product of a mixed set of influences from logic, mathematics and physics, including Hilbert's formalism and Frege's logicism, Russell's conception of an ideally perfect logical language, and Einstein's analysis of simultaneity and of geometry.

From the general philosophical point of view, however, the standard conception was the result of a complex work of analysis, consistently directed towards a technically rigorous formulation of the fundamental theses of the *wissenschaftliche Weltauffassung*, the scientific world-conception. "The scientific world-conception – Neurath, Hahn and Carnap declared in the famous 1929 Manifesto – knows no unconditionally valid knowledge derived from pure reason, no 'synthetic judgments a priori' of the kind that lie at the basis of Kantian epistemology and even more of all pre- and post-Kantian ontology and metaphysics. [...] It is precisely in the rejection of the possibility of synthetic knowledge a priori that the basic thesis of modern empiricism lies. The scientific world-conception knows only empirical statements about things of all kinds, and analytic statements of logic and mathematics".[1]

If this programme, largely shared by the other members of the movement, especially Schlick and Reichenbach, was to be successful, two fundamental results had to be achieved.

In the first place, a philosophical account was needed of the apparently universal and necessary components of knowledge (especially logical and mathematical truths). This account had to rely exclusively upon the distinction between analytic and synthetic sentences, so as to

[1] Neurath, Hahn, Carnap 1973: 308.

avoid any reliance upon Kant's conception of the synthetic a priori, or upon Husserl's conception of the material a priori.

Secondly, it was necessary to elaborate a criterion of meaning, or at least of cognitive meaningfulness, which would make it possible to distinguish between empirical, cognitively meaningful sentences, and non-empirical, cognitively non-meaningful sentences. This was required in order to purify scientific discourse of all metaphysical contamination, by showing that metaphysical sentences are really pseudo-statements without meaning.

This second task was entrusted to the verifiability principle (the 'empiricist criterion of cognitive meaningfulness'): the meaning of a sentence is the method of its empirical verification. The standard conception developed largely out of the attempt to give increasingly satisfying formulations of the verifiability principle. The adequacy of the suggested formulations was assessed on the basis of their capacity to exclude paradigmatic examples of metaphysical speculation from cognitively meaningful discourse, without compromising the meaningfulness of the essential elements of advanced scientific knowledge, that is, of metric and dispositional concepts, universal scientific laws, theoretical terms and statements. Paradigmatic metaphysical speculations were not – as Popper would have us believe – the Presocratic cosmologies, but rather Driesch's vitalism, speculative constructions in the style of Bergson and Heidegger, the systems of nineteenth century idealism, the epistemological opposition between realism, idealism and phenomenalism.

1.2. THE CONTRAST BETWEEN CARNAP AND HEMPEL

I will not dwell on the internal and external discussions which accompanied the liberalisation of neoempiricism, nor will I illustrate in detail the logical and epistemological features of the definitive version of the standard conception. It will be sufficient to recall the important contrast that emerged at a certain point between Carnap and Hempel.

After the fifties, only Carnap remained faithful to the two fundamental aims of the scientific world-conception: the formulation of an empiricist requirement of cognitive meaningfulness, and the distinction between analytic and synthetic statements. Carnap, embracing Duhem's holism on experimental testing, and developing to its extreme consequences a line of thought commenced in the *Logische Syntax der Sprache* (1934), in the end abandoned completely the reductionist idea of the empirical testability of isolated sentences. But he did not give up talking about analyticity and the empirical meaningfulness of sentences. He maintained that holism requires only that analyticity and empirical meaningfulness be seen as *relative* to an appropriate system made up of: (a) a language, with its syntactic and semantic rules; and (b) a set of theoretical principles and interpretative rules (bilateral reduction sentences). Carnap, responding to Quine's and Hempel's objections, formulated an ingenious empiricist criterion of cognitive meaningfulness, uncompromised by reductionism. Moreover, a skilful use of the so-called 'Ramsey sentence' enabled him to distinguish between meaning postulates and empirical postulates, between that which pertains to the 'specification of meanings' and that which pertains to 'the assertion of empirical facts'.

Carnap believed right up to the end that he had succeeded in constructing the tools needed to rigorously purify scientific discourse of all metaphysical contamination. In his writings, the standard conception continued to present itself as the definitional explication of the notions of analyticity and cognitive meaningfulness, and therefore as the fulfilment of the scientific world-conception. Carnap's explanation of analyticity and cognitive meaningfulness consisted in specifying the logical, structural conditions which a theory and the sentences that made it up should satisfy if they were to be considered cognitively meaningful and not metaphysical. His proposal was to be seen both as a logical reconstruction of the notion of scientific theory and, at the same time, as the philosophical specification of the structural characteristics of an ideally perfect language, in which it would be possible to express only meaningful discourse. Carnap formulated the logical rules concerning the formation of sentences and their (purely tautological) deductive transformations, as well as a semantic requirement able to discriminate between meaningful and non-meaningful expressions.

Relying upon his general conception of explication, Carnap presented this logical reconstruction of the notion of scientific theory as something in between the normative ideal and the descriptive model. On the one hand, his reconstruction prescribed the ideal conditions which had to be satisfied by a theory if it were to be considered scientific. On the other hand, it was structured so as to present the salient characteristics of certain existing theories believed to be paradigmatically scientific and meaningful.

However, despite 'liberalisation' and the relativisation of the fundamental concepts of meaningfulness and analyticity, Carnap's project was unsuccessful. Right from the beginning, the standard conception was severely criticised in virtually all its characteristic aspects. The whole story – by now a classical debate in the history of philosophy – is told in Suppe's famous reconstruction.[1] Among the first critics were Achinstein and Putnam. As well as starting the attack against the distinction between theoretical and observational terms and entities, they criticised the notion of partial interpretation on which the whole standard conception rests. In the following years, criticisms inspired by the 'later' Wittgenstein, by Polanyi, Hanson, Kuhn and Feyerabend, were raised against the other principal characteristics of the standard conception, viz. the assimilation of scientific theories to linguistic entities, that is, to complexes of sentences presented in axiomatic-deductive form; the

[1] Suppe 1977a and 1977b; see also Suppe 1989, p. I, 'From the received view to the semantic conception' (pp. 1-77).

excessive importance attributed to formalisation and the corresponding underestimation of the function of models in the development of empirical sciences; the two-level vision of scientific discourse, with the dichotomy between theoretical and observational language; the way of looking at the rules of correspondence; the excessive concessions to extensionalism; the possibility of tracing a clear boundary between meaningful and non-meaningful expressions, and between analytic and synthetic sentences.

Since I intend to look at general philosophical problems, I will not dwell on these strictly technical criticisms. It is sufficient to say that, to a large extent, they hit the mark and are by now generally accepted. The philosophically relevant point is that these criticisms of the standard conception affected also the fundamental theses of logical positivism, to the point that even Hempel, one of the major exponents of the movement, distanced himself from Carnap's unremitting attempts to salvage them.

Hempel has come progressively closer to Quine and to the new philosophers of science, moving beyond the fundamental principles of the *wissenschaftliche Weltauffassung*. He has questioned the epistemological relevance of the notion of analyticity. He has dropped the empiricist criterion of meaningfulness, denying the existence of a *clear* boundary between metaphysics and the more theoretical components of scientific discourse. He suggests instead a less binding distinction of degree between complexes of sentences with more or less empirical content. In recent years, he has considered the limits of a purely hypothetical-deductive reconstruction of theories, and he has reinterpreted the dichotomy between theoretical and observational vocabulary as a distinction between languages of new theories and languages of theories which are already in use and understood.

The dichotomy between theoretical and observational vocabulary was originally introduced by the neopositivists to clarify how theories receive a partial empirical interpretation by means of their direct and / or indirect connection with observational statements. Consequently, the crisis of the standard model has made problematic not only the desired semantic demarcation between science and metaphysics, but also the empirical value of science or, in other words, the relevance of experience in the validation and refutation of our hypotheses and theories.

For an epistemological view inspired by empiricism, the outcome could hardly seem more disastrous!

1.3. THE CRISIS OF NEOPOSITIVISM AND THE SURVIVAL OF THE POSITIVE SPIRIT IN THE 'CONVERSATION OF MANKIND'

Therefore, there are clear and strong reasons why the crisis of the standard conception is considered by many contemporary philosophers not merely as the end of a model of scientific theories, but as the end of the whole general philosophical perspective upon which this model rested. This helps explain the present proliferation of anti-positivist and anti-empiricist philosophies, even within philosophical circles deeply permeated with logic, analysis of language and epistemology. The cultural climate is not too dissimilar to that which followed the polemics against mechanistic physics at the beginning of the century, when the crisis of mechanism was interpreted by some as the 'bankruptcy of science' in general. The waning of neopositivism has coincided with the widespread conviction that 'positive' philosophies are remnants from history with no future ahead of them. To sum up, it is widely believed that the developments in philosophy of science over the last four decades have marked the doom of positivism, the demise of an intellectual enterprise which had its beginnings in the first half of the nineteenth century, and has now met its end.

I find this view unconvincing. For reasons that I have been expounding for almost twenty years, I am inclined to think that the situation for empiricist and positivist philosophies is not as desperate as many seem to believe. On the other hand, I am persuaded that the survival of these philosophies is conditional upon deep changes being made to the old and new empiricisms and positivisms of the past. I would like to support this 'outdated' conviction of mine by developing a secular and detranscendentalised *positive* philosophy which will not be vulnerable to the criticisms raised against logical positivism.

Despite the crisis of neopositivism, I think it is still possible to defend the role of the supporters of the positive spirit in the 'conversation of mankind'. I have become increasingly convinced of this for two reasons.

(1) There are good reasons to suppose that the criticisms of the basic theses of the standard conception do not really affect the fundamental principles of the positive spirit, but rather the particular interpretation that the logical empiricists (or better, some of the logical empiricists, especially Carnap) gave of these fundamental principles (section 3, below).

(2) On close scrutiny, some of the salient characteristics of the present philosophical situation seem to support, rather than discourage, the aspirations towards a positive philosophical point of view (section 4, below).

But, before I can illustrate all these reasons (in sections 3 and 4, below), I must briefly consider the great guiding ideas of positive philosophy, which the logical empiricists tried to reformulate and strengthen.

2. THE POSITIVE SPIRIT: RELATIVISM, EMPIRICISM, ANTI-METAPHYSICAL OBJECTIVISM

I believe that these ideas have been clearly indicated in the works on the history of logical empiricism written by Philipp Frank, a participant in that movement, and by Victor Kraft, a scholar who was strongly influenced by it.

Kraft focused on a salient aspect of positivism, characterising it as "a fundamental philosophical attitude, a position of the spirit", rather than as a somehow closed "unitary system of thought", which can be identified with a specific philosophy (e.g., Comte's). (Incidentally, Carnap repeatedly said something analogous about Neurath's physicalism.)[1] This position of the spirit means, in the first place, a critical stance towards knowledge considered paradigmatically as "'positive' knowledge", namely, "the specialised sciences", in opposition to an alleged "transcendent" metaphysical knowledge.[2]

Frank explained the fundamental features of this position of the spirit quoting the words with which Abel Rey, in 1907, illustrated Poincaré's conventionalism. Rey describes Poincaré's philosophy as a "new positivism", a marked improvement on Comte and Mill's positive philosophy. This passage is particularly significant, because Frank resorted to it to illustrate the origins and the spirit of the philosophy of the Circles of Vienna and Berlin. It is true that, if we look at Poincaré's thought as a whole, there is a strong contrast between a tendency towards accentuated forms of conventionalism and subjectivism on the one hand and a tendency in the direction of positivism and objectivism on the other. But, setting aside this subtle exegetic point (already noted by Duhem),[3] we can take Rey's words as the incisive indication of those aspects of Poincaré's thought which can be considered to be the expression of the positive spirit.

According to Rey, Poincaré's "new positivism" preserves "the great directive ideas" of Comte, Taine, Mill with respect to "the relativism and empiricism of our knowledge". At the same time, it differs from traditional positivism in that "it stresses ... the idea of experimental categories which are required by science as a central necessity. [...] What was lacking in Comte's or Mill's positivism ... was their ... failure to have established in a new form a theory of categories. Objective experience is not something which is outside and independent of our minds. Objective experience and mind are functions of each other, imply each other, and exist by virtue of each other. To say that the relations between physical objects derive from the nature of those objects and to say that these relations are constructed by our minds are two artificial theories". Poincaré realised that "[o]ur experience is a system, a relation of relations. The relation is the given".[4] His new positivism had the historical merit of shifting traditional positive philosophy towards rationalism and conventionalist relativism, without giving up the idea of the empirical and objective character of scientific knowledge. Frank maintained that neopositivism moved in the same direction. It too tried

---

[1] See, for example, Carnap 1963a: 51; see also Neurath 1935 8

[2] See Kraft 1961: 817f.

[3] Duhem 1962b: 328.

[4] Rey 1907: 392ff, cited in Frank 1975· 9f

to improve nineteenth century positive philosophy, developing its rationalist component along with its empiricist one, in the light of the subsequent great developments in foundational mathematical research, symbolic logic, relativistic physics and quantum mechanics. It is significant that Neurath, citing Gregorius Itelson, used the expression 'empirical rationalism' as well as 'logical empiricism'.[1]

Despite the emphasis on the subjectivist and rationalist aspects of science (an emphasis which was not properly understood for a long time), the logical empiricists agreed with the old positivists in underlining the objective value of knowledge. Objectivity was upheld – Kraft wrote – alongside with the relative character of scientific statements, rejecting the possibility of all a priori knowledge of reality and all metaphysics – where metaphysics is intended not as "speculative formation" of hypotheses "on the basis of experience", but as absolute knowledge which claims to be "transcendent" with regard to experience.[2] These too are typical aspects of positive philosophy, to be found in both major and minor representatives of nineteenth century positivism: from Comte and Spencer to Mill and Herschel, from Tyndall and Du Bois-Reymond to Ampère and Littré. Though Comte, the father of positivism, admits (as Mach did later) the influence of theory on observation, thereby avowing the limits of narrowly empiricist conceptions of knowledge, nevertheless he repeatedly maintains that the evolution of the human spirit leads gradually to the recognition of the impossibility of attaining absolute notions and foundations. Using similar expressions, Poincaré excludes from scientific knowledge all explanations of natural phenomena which claim to go beyond the search for purely relational connections, in the vain attempt to reveal the essential properties of phenomena, or, as Mill used to say, their "ultimate causes". The attempts to discover the "essence" of electricity and magnetism are for him "the affair of metaphysicians". As did Mach, he repeatedly accuses metaphysics of being idle, meaningless, impracticable.[3]

But neither the insistence on the conventional aspects of science, nor the rejection of the idea that science yields knowledge of the hidden reality behind the phenomena, threatened the objectivity of scientific knowledge. Against Le Roy's extreme conventionalism and instrumentalism, Poincaré argued that we cannot account for the practical success of science unless we accept that it makes us know something of reality.[4] As a philosopher has repeated recently, underlining yet again the dangers of extreme forms of pragmatism, "[a]s tools, scientific theories can indicate to us something about reality – if only the mere fact that some aspects of it are amenable to treatment of a certain sort with that kind of

---

[1] Neurath 1935· 44.

[2] Kraft 1961: 818.

[3] See Poincaré 1946b: 28, 29f, 140f (for aether as a metaphysician's business, p. 174); 1946c: 349f, 352

[4] Poincaré 1946c: p. III, ch X, sec. 2, 'Science, Rule of Action' (pp. 323-5)

tool".[1] The point is, for Poincaré, that this something is not "the things themselves [...] as the naïve dogmatists think, but only the relations between things. Outside of these relations there is no knowable reality".[2]

The history of science teaches us precisely this. It shows us that scientific theories grow old and decay very rapidly. But, if we look more carefully, we see that what really die are their claims to describe the reality hidden behind the phenomena, and not the complex of empirical relationships underlying the theories. Once a theory has helped us discover a real relationship between things, this relationship will be found, under a new aspect, even in later theoretical constructions. Goethe once talked about hypotheses as "scaffolds which are erected in front of a building and removed when the building is completed. They are indispensable to the worker; but he must not mistake the scaffolding for the building": "[t]he constancy of phenomena alone is important; what we think about them is quite immaterial". Frank rightly connected these words with Mach's assertion that "[t]he aim of natural science is to obtain connections among phenomena. Theories, however, are like withered leaves, which drop off after having enabled the organism of science to breathe for a time".[3]

It is now clear which are the fundamental principles of the positive spirit, inherited by the Vienna and Berlin Circles from the old nineteenth century positivism, through the mediation of Poincaré. These principles are: (1) the rejection of any allegedly absolute knowledge which transcends the limits of the positive sciences (i.e., metaphysics); and, (2) the conviction that the knowledge offered by the usual scientific disciplines is characterised by three somewhat contrasting features: it is relative, empirical and objective. I will condense all this by talking of relativism, empiricism and anti-metaphysical objectivism.

In order to evaluate the impact of the crisis of the standard conception on these fundamental principles of positive philosophy, we must look at the peculiar way in which these principles were reformulated in the philosophy of the Vienna and of the Berlin Circles.

3. THE CRISIS OF NEOPOSITIVISM AS CRISIS OF THE LINGUISTIC TURN

Today we know very well that neopositivism was not a monolithic movement. Internally, it had deep divergences and major disputes. Reichenbach accused of dogmatism the non-probabilistic and anti-realist verificationism that characterised the initial phases of the Vienna Circle. Neurath, in constant conflict with Schlick on the nature of philosophy and of truth, expressed ideas which are now considered antici-

---

[1] Johnson 1992: 239.

[2] Poincaré 1946b: 28.

[3] From Goethe, *Maxims and Reflections*, maxims n. 1222, 1229, and from Mach 1911, cited in Frank 1975: 62f.

pations of the criticisms raised by the new philosophers of science against the epistemology of Carnap, Hempel and Popper. However, once we have allowed for many important differences, I believe we can say that the principal tools with which the logical empiricists (especially Carnap) tried to realise the ideals of the positive spirit, are tied to the 'linguistic turn' of philosophy. The linguistic turn, connected with the logical analysis of language, was described in the 1929 Manifesto as the main characteristic distinguishing the new positivism from nineteenth century positivism, which was instead biologically and psychologically oriented.

We can say that neopositivism gives a sort of 'logical-linguistic clothing' to the ideals of the positive spirit. Neopositivism is the attempt to realise these ideals using the tools of twentieth century analytic philosophy of language, especially the verifiability principle (variously interpreted) and the linguistic theory of the a priori. On the other hand, it is the heritage of the positivist tradition that sets logical positivism apart from the other two main variants of twentieth century analytic philosophy, namely, logical atomism (from which it differs because of its anti-metaphysical attitude), and the analysis of ordinary language (from which it differs because of its attention to the logical and linguistic structure of the positive sciences).

It is somewhat trite to point out that the substitution of the old biological and psychological approach by the logical analysis of language – a substitution which is not the same in all the representatives of the Vienna and the Berlin Circles – strongly conditioned the new philosophical formulations of the logical positivists, starting with the kind of criticism they raised against metaphysics. This is the case for Schlick, who denied that philosophy can make meaningful metaphysical statements and assigned to philosophy the task of elucidating the meaning of sentences. This is the case for the unorthodox Neurath, who, though he rejected as metaphysical the ideas of the *Tractatus* on the logical form of language, recognised "the 'linguistic' origin of many philosophical problems".[1]

This is especially the case for Carnap, who contributed more than anyone else to the diffusion of the standard image of neopositivism. In Carnap's work, criticism of metaphysics and defence of the scientific world-conception are not simply a rejection of the various forms of absolutism (connected to nineteenth century idealism, to metaphysical realism and to the absolutisation of the data of experience) attacked by Schlick in 'Positivismus und Realismus'. Rather, Carnap's criticism of metaphysical constructions such as Heidegger's takes the form of *Überwindung der Metaphysik durch logische Analyse der Sprache* [elimination of metaphysics through logical analysis of language]. Philosophy is made to survive as an analysis-construction (explication) of the "forms of language" which should be used to reformulate scientific discourse in a philosophically satisfactory way.

---

[1] Neurath 1935: 12.

If this is the situation, it is legitimate to ask whether it is the guiding ideas of positive philosophy, or whether it is, in a more limited way, the logical-linguistic form in which these ideas were recast by the neopositivists, which has not survived the criticisms of the standard conception. In order to answer this question, I will start from the relations between the verifiability principle and the rejection of metaphysics (section 3.1., below). Then I will turn to the linguistic conception of the a priori (section 3.2., below) and to the reconciliation between relativity and objectivity of knowledge (section 3.3., below).[1]

3.1. VERIFIABILITY AND ANTI-METAPHYSICS AS ANTI-ABSOLUTISM

A. The meaninglessness of metaphysics

From an historical point of view, the neopositivists presented their anti-metaphysical polemic as a revival of Mach's programme of purifying science of the remnants of purely speculative thought. In Mach (as in the best tradition of modern science) this project is connected with the golden rule of the scientific method, according to which one should not introduce aspects of reality that do not have an empirical justification.

The verifiability principle gives this rule a linguistic value, incorporating it in the linguistic turn of philosophy, which was intended to replace the analysis of ideas and of the faculties of the human mind which characterised Locke's and Hume's old empiricism. This linguistic version is rooted in some important aspects of late nineteenth century and early twentieth century scientific and methodological thought (in particular, in some ideas of Peirce and of Poincaré). An especially important influence, as the neoempiricists did not tire of emphasising, was Einstein's analysis of simultaneity. It is certain, however, that the 'verification theory of meaning' was originally formulated by the two Viennese Schlick and Carnap, under the influence of ideas developed by Russell and Wittgenstein in the years around 1920. In that first version, the verification theory of meaning comprised many of the notions that were to be inherited later by the standard conception (for example: the distinction between atomic and molecular statements, and the extensionalism of the principle of truth-functional composition), as well as Wittgenstein's idea that the understanding of a statement can be traced back to knowing what is the case when the statement is true.

Connecting these ideas to the empiricist and sensationalist aspects of Hume's and Mach's philosophy, the neopositivists maintained that statements which are in principle unverifiable are not simply empty, useless or idle, but without meaning. They condemned as meaningless and unintelligible all non-empirically reinterpretable metaphysical statements. In metaphysical speculation, we seem to understand what we are talking about, but really we do not understand it at all. Even in the cases in which the rules of grammatical formation of discourse are re-

---

[1] See Schlick 1979c, especially pp. 272ff; Carnap 1966.

spected – and in some cases not even this happens – the conditions for the semantic intelligibility of what is said are lost.

B. Liberalisation: the cognitive *meaninglessness of metaphysics*

This radical anti-metaphysical stance gave rise immediately to many technical problems and many fundamental philosophical objections. Among the technical problems, it is enough to recall the difficulties concerning the way theoretical statements and empirical observations were to be connected, or the way of interpreting the empirical basis. And the contrasts between phenomenalism and physicalism, or the philosophical clarification of the notion of observation, seemed to many thinkers the renewal of old quasi-metaphysical controversies. However, the most famous philosophical objection sprung from the application of the verifiability principle to itself. This was intended to question the empirical verifiability, and therefore the meaningfulness, of the verifiability principle. This criticism cast doubt on the very possibility of an empiricist philosophy, based on verifiability as a criterion of meaning.

In order to make their position coherent, the logical empiricists made use of a whole series of distinctions and limitations, substantially attenuating their initial claims. Carnap, in particular, restricted the scope of the verifiability principle characterising it as a criterion not of meaning, but of meaningfulness and, on top of that, not of meaningfulness in general, but of *cognitive* meaningfulness. This was equivalent to recognising:

(1) It is not possible for the meaning of the elements of apophantic (i.e., declarative) discourse to be completely exhausted in the conditions of their empirical applicability. That is to say, the elements of predicative discourse can have a surplus of meaning in addition to that which is empirically determinable.

(2) Statements empty of cognitive value "often have other meaning components, e.g., emotive or motivative ones, which, although not cognitive, may have strong psychological effects".[1] We must remember that Carnap had always spoken of metaphysicians as would-be musicians who, lacking musical talent, make use of improper theoretical formulations, in order to convey emotional meanings, whose adequate expressive form would be artistic. Incidentally, I am convinced that a lexicographical analysis of Carnap's writings would easily show a certain overlap between meaninglessness and absence of scientific and cognitive import.

Anyhow, in the final version of the standard conception, Carnap employed two different conceptions of meaning.

(1) As Putnam incisively put it, he used "the customary (linguistic) conception"[2] when the task was the evaluation of the grammatical form of expressions, the synonymy of words, the analyticity of sentences. The evaluation depended upon the syntactic and semantic

[1] Carnap 1963a: 45f

[2] Putnam 1975c: 124.

rules (including meaning postulates) of the observational and theoretical language in which a theory is formulated.

(2) Carnap used the verifiability conception of meaning when the task was the evaluation of the meaningfulness of an expression. In this case, the criterion for assessing the empirical status of a theory and of its components (terms and sentences), was the existence of empirically testable consequences.

This dualism seriously weakened the charge of meaninglessness originally directed against metaphysics. This accusation could no longer be intended in the strong sense of linguistic meaninglessness. It was reduced to lack of empirical value, and therefore, to be generous, to lack of scientific and cognitive value.

To this it must be added that, in order to overcome the difficulty raised by the verifiability (and therefore by the meaningfulness) of the verifiability principle, Carnap and Hempel were induced to ascribe to it a definitional value and not a descriptive one. It is therefore understandable how the neopositivist criticism of metaphysics seemed to many thinkers to be the result of a 'persuasive redefinition' of the concept of meaning. To borrow Putnam's words once again, "[i]f to call metaphysical propositions 'meaningless' were only to assert that these propositions are empirically untestable, it would be harmless (the metaphysicians always *said* that their assertions were neither empirically testable nor tautologies); but, of course, it is not harmless, because the Positivist hopes that we will accept his redefinition of the term 'meaning', *while retaining the pejorative connotations of being 'meaningless' in the customary (linguistic) sense, i.e. being literally without sense*".[1]

As if all this were not enough, the formulation of a satisfactory criterion of cognitive meaningfulness ran into such difficulties that most thinkers were induced to follow Hempel in abandoning the enterprise.

C. Anti-metaphysics as anti-absolutism

Personally, I remain of the view that an adequate theory of the learning and functioning of language cannot sever all ties between the meaning of our descriptive expressions and the conditions of their empirical applicability. Consequently, I think that, in addressing the problem of meaningfulness, one must retain a role for the relation with experience. But, even granting the complete failure of the verifiability principle, would it be justified to infer from that the failure of anti-metaphysics? I doubt it.

It is clear from the writings of Mach, the father of the Vienna Circle (which was also called *Verein Ernst Mach*) that the anti-metaphysical stance meant essentially a critical stance, *anti-absolutism* (this is particularly clear in the *Mechanik*). Anti-metaphysics meant, in the first place, rejection of all those concepts, principles, entities and processes (for instance, the notions of absolute space, time and motion, or the hy-

---

[1] Putnam 1975c: 122f.

postatisation of physical laws as structures of reality in itself), which aspired to a validity that was more than merely relative, let's say, to the specification of a system of reference, or to the accidental historical circumstances of their formulation. If we stay with this original formulation of anti-metaphysics (typical of the positivist tradition), it is difficult to infer from the failure of the verifiability principle, the failure of the critique of metaphysical absolutism. This seems all the more true if we consider the strong relativist components of the developments in the philosophy of science which led Hempel to abandon, and Carnap to relativise, the empiricist criterion of meaningfulness.

Lacking evidence to the contrary, we can conclude that these developments do not undermine the critique of metaphysics as such, but rather the particular logical-linguistic form that this critique was given by the neopositivist reliance on the verification theory of meaning.

3.2. ANALYTICITY AND EMPIRICISM

A. Neoempiricism: the linguistic theory of the a priori

I believe that something similar can be said about the objections to the neopositivist conception of analyticity, the a priori, and the principles of mathematics and logic. This conception brings together Frege's and Russell's thesis of the reducibility of mathematics to logic (logicism), and Wittgenstein's idea of the tautological character of the propositional calculus. Correspondingly, many objections hinge on the difficulty of justifying logicism and proving the tautological character of the principles of the quantification theory and of the theory of classes, for example, of the axioms of comprehension or of infinity.

But my concern is only with the criticisms relating to the fundamental philosophical motivations of neopositivism. These motivations were of two kinds.

(1) The logical empiricists agreed with the substance of the criticisms directed by Bolzano, Frege and Husserl against empiricist and psychologistic theories of logic and mathematics. Accordingly, they were unwilling to consider logical and mathematical principles as contingently true statements.

(2) They were firmly set against anti-empiricist conceptions such as Kant's theory of synthetic a priori judgements and Husserl's theory of the material a priori (explicitly criticised by Schlick in 'Gibt es ein materiales Apriori?').

The linguistic conception helped on both counts: the linguistic nature of a priori truths insured they were apodictic and independent of experience, denying at the same time that they had any ontological or cognitive content. Moreover, on this conception, the truths of logic and mathematics can be equated neither to descriptions of concrete thought processes, nor to inductively founded empirical generalisations; instead, they are assertions in their own right, whose a priori validity depends on the linguistic rules governing the use of logical and mathematical symbols.

In 1933, Hans Hahn gave a particularly clear formulation of these theses in the booklet *Logik, Mathematik und Naturerkenntnis*. In it, the Viennese mathematician disputed the traditional conception of logical and mathematical principles as descriptions of the universal properties of things. Denying them any ontological import, he maintained that they have a linguistic basis: they do not concern things, but the structural features of the language we use to talk about things. The certainty and universality of logic and mathematics hinge precisely on the fact that they do not assert anything about any particular object. All one needs to do in order to see this is to consider some simple exemplifications of the laws of contradiction and excluded middle: 'No object is both red and not red' and 'Every object is either red or not red'. "These two propositions [...] say nothing at all about these or any other objects; nor do I learn from them whether an individual object is red, whether it is not red, what colour it has, or anything else: these propositions merely stipulate something about the kind of way we want to assign the designations 'red' and 'not red' to objects, i.e., they stipulate something about *how we want to talk about objects*".[1] In this way, logical and mathematical statements ended up in the same category with trivially analytic statements of the kind made famous by Quine, 'Bachelors are unmarried men'.

Moreover, the logical empiricists thought they could reduce the more general, abstract principles of empirical science, for example, the principles of causality or of induction, to methodological requirements, or (as I will specify shortly) to linguistic conventions. The conjunction of this idea with the linguistic theory of the a priori made it possible for the logical empiricists to classify all meaningful apophantic sentences as either analytic or synthetic statements. This classification was considered a linguistic improvement of analogous traditional mentalistic distinctions, such as Hume's distinction between relations of ideas and matters of fact, and Kant's distinction between analytic and synthetic judgements. The result was the denial, in keeping with the scientific world-conception, of synthetic a priori truths, and the consequent identification of analytic statements with a priori necessary truths, and of synthetic statements with a posteriori contingent truths.

B. Quine's empiricist critique of the linguistic theory of the a priori

I chose Hahn's passage because it is particularly suited to highlight the problems on which the linguistic conception was eventually to run aground.

(1) Reading Hahn with the advantage of hindsight, it is clear that his solution would bring about the desired defeat of traditional ontology and mentalism, only if it were accompanied by a theory of language and meaning, devoid of ontological commitments to abstract or mental

---

[1] Hahn 1987: 30.

entities. Quine's later criticisms of intensional and intentional notions exposed the difficulty of eliminating from semantics all forms of Platonism and mentalism.

(2) In the case of logical truths, Hahn's formulation displays another aspect of the issue – an aspect somehow already implicit in the *Tractatus*. Let us grant the legitimacy of the reliance on the notion of meaning. Let us even grant the impossibility of denying sentences, such as 'It is either raining or it is not raining', or 'An object is either red or it is not red', without at the same time modifying the meaning of the logical words 'or' and 'not'. The question is: in what sense can we say that sentences of this kind do not say anything about objects or about reality? To be sure, they do not tell us that a given object is red or that it is not red, nor that it is raining, or not raining. Nevertheless, they do tell us that reality is such that an object is either red or not red, or that it is either raining or not raining. That is, they tell us that certain characteristics of reality can be described by means of words that have the meanings we habitually associate with 'or' and 'not'. As Quine effectively put it: "Consider [...] the logical truth 'Everything is self-identical', or '$(x)\ (x = x)$'. We *can* say that it depends for its truth on traits of the language (specifically on the usage of '='), and not on traits of its subject matter; but we can also say, alternatively, that it depends on an obvious trait, viz., self-identity, of its subject-matter, viz., everything".[1]

In conclusion, even accepting (as Quine would not) the notion of meaning, and the connection between logic and the structural aspects of our language, it is difficult to affirm the ontological and descriptive emptiness of logic. In so far as we assert them with reference to objects, logical truths entail also the description of the more general traits of reality. Personally, I am inclined to believe that this 'ambivalence' contains the solution to the thorny problem of the epistemological status of logic and mathematics. Moreover, I do not see how it is possible to set the truths of logic and mathematics apart from the more general and abstract statements of the empirical sciences by claiming that logic and mathematics say nothing about the world. This claim is clearly falsified by the well known connections between logic and quantum physics.

The attempt to found logical and mathematical truths on direct empirical induction was mistaken, since it is clearly impossible to justify the validity of 5+7=12 on the basis of the repeated observation, let's say, that, putting five rabbits and seven rabbits together in a pen, we end up with twelve rabbits. But the linguistic theory of logic and mathematics failed to exclude completely their descriptive import with regard to being or reality. The linguistic theory was not capable of precluding the alternative thesis, suggested some decades later by Quine, who appealed to the diffuse character of empirical content, in order to assimilate logic and mathematics to the more abstract theoretical parts of the natural sciences.

---

[1] Quine 1977: 113

As Quine says in *Philosophy of Logic*, "[m]athematics and logic are supported by observation", not in the sense of their being inductively extracted from specific empirically given cases, but "only in the indirect way" in which even "the most general and systematic aspects of natural science, farthest from observation" are supported by experience; "namely, as participating in an organized whole which, way up at its empirical edges, squares with observation". From this point of view, the thesis of the "empirical character of logic and mathematics" does not have greater epistemological significance than the thesis of the "unempirical character of theoretical physics". What counts is the "kinship" between the assertions of these two disciplines, that is, the gradual character of the passage from statements closer to experience to statements farther from experience, but endowed nevertheless with empirical content.[1]

Therefore, Quine's epistemology – which has proved itself to be one of the most convincing alternatives to the neopositivist conception – does not entail the unconditional surrender of empiricism. On the contrary: its holism and gradualism confer upon logic and mathematics an empirical, hypothetical and descriptive character similar to the generally acknowledged character of the theoretical principles of the empirical sciences. Nor does this entail the return to unacceptable forms of logical and mathematical inductivism, or to unjustified metaphysical interpretations founded (as was Russell's logical atomism) on the presupposition of an isomorphism between the logical structure of the ideally perfect language and the ultimate structure of reality.

We can conclude that the crisis of neopositivism concerned a particular, linguistically oriented, way of reconciling empiricism with logic and mathematics; it has not been proved that this entails the rejection, rather than the strengthening, of the empiricist perspective.

3.3. LINGUISTIC CONVENTIONALISM, RELATIVISM, AND OBJECTIVISM

A. The subjectivist component of neopositivist epistemology: the weak rejection of the a priori

In the case of the reconciliation between relativism and objectivism too, I believe we should distinguish between the impossibility of attaining the reconciliation, and the inadequacy of the kind of linguistic solution to the problem suggested by the neopositivists.

This issue is connected to what is, in my view, one of the most important, and most misunderstood, aspects of logical empiricism. Since the years around Enriques' contribution to the Paris Congress of Scientific Philosophy (1935), it has been widely believed that the members of the Circles of Vienna and Berlin underestimated the role of the "activity of mind" in the construction of the "scientific synthesis" (this was Enriques' criticism of the philosophy of the Vienna and Berlin

[1] Quine 1986: 100.

Circles).[1] A closer reading of the texts, now abundantly confirmed by archival research, has shown the neopositivists' full awareness of the role played in scientific knowledge not only by logic and mathematics, but also by other "forms of thinking" (as Frank called them)[2] that rely on the exercise of a spontaneous inventive capacity, largely autonomous of experience.[3]

The neopositivist rejection of Kantian transcendentalism meant only the denial that this spontaneous intellectual activity yields apodictically certain synthetic a priori principles (weak rejection of the a priori).

Even in its more reductionist phases, the neopositivist critique of apriorism never excluded the role of abstract conceptual structures, such as the notion of causality, the principle of induction, the geochronometric assumptions at the basis of measurement. The logical empiricists only changed the interpretation of these conceptual structures, considering them, according to the occasions, as extremely general hypotheses; constitutive principles globally dependent on experience; methodological requirements which cannot be subjected to the verifiability principle; and, lastly, as linguistic conventions empty of factual content.

B. Linguistic conventionality of the a priori, the pillar of neopositivist objectivism

Having recognised the subjectivist component of neopositivist epistemology, we must be careful not to interpret this component as entailing a radically subjectivist and relativist epistemology. The neopositivists believed – in a way that is in some respects reminiscent of Poincaré – that the dependence of knowledge on subjective conventional components could be reconciled with the objectivity of scientific statements.

A clear example of this is Reichenbach's interpretation of the "relativity of geometry" with respect to the coordinative definition of congruence: "[*p*]*roperties of reality are discovered only by a combination of the results of measurement with the underlying coordinative definition.*" "*The objective character of the physical statement is thus shifted to a statement about relations*. A statement about the boiling point of water is no longer regarded as an absolute statement, but as a statement about a relation between the boiling water and the length of the column of mercury. There exists a similar objective statement about the geometry of real space: *it is a statement about a relation between the universe and rigid rods*. [...] In this sense we are permitted to speak of *physical geometry*. The description of nature is not stripped of arbitrariness by naive absolutism, but only by recognition and formulation of the points of arbitrariness. The only path to objective knowledge

---

[1] Enriques 1936: 26

[2] Frank 1975: 7ff.

[3] For a more detailed justification of the historical component of my theoretical approach, see Parrini 1980b; 1983; 1987, 1994b, 1993, 1994c; 199+a.

leads through conscious awareness of the role that subjectivity plays in our methods of research".[1]

The neopositivist combination of relativism and objectivism was founded on the conviction that the "points of arbitrariness" in scientific knowledge were ultimately *linguistic* conventions, empty of factual content. This conviction – already present in Poincaré – was entirely compatible with the reductionist formulations of the verifiability principle and of the linguistic conception, according to which it is possible to talk about the empirical content of an isolated sentence.

This tangle of interconnections, which I have tried to unravel elsewhere,[2] shows that Quine was not completely wrong when, in criticising the "two dogmas of empiricism" (analyticity and reductionism), he maintained that they are "at root identical".[3] It is true that, from a theoretical point of view, Quine's thesis has proved itself mistaken, since it has been repeatedly shown that it is possible to accept analyticity whilst dropping reductionism. But, keeping in mind the historical context in which Quine's thesis was set forth, we can try to reformulate it by saying that for a long time, in the neoempiricist theory of knowledge, the two dogmas have supported each other: it was their conjunction which made it possible to reduce important "forms of thinking" to linguistic conventions, thus allowing the reconciliation of relativism and objectivism which we have seen in Reichenbach's passage on geometry.[4] As we shall see, the problems become harder when we recognise that conventionalism, and therefore subjectivism and relativism, concern theoretical and even observational, as well as linguistic, components of scientific discourse.

However, it has not been proved that the impossibility of basing the reconciliation between relativism and objectivism on the notion of *definitional* convention forces us to choose between metaphysical objectivism and radical subjectivist relativism, of the kind epitomised by the unpalatable slogan of the 'loss of the world'. Personally, I do not think that if we give up absolutism we have to resign ourselves to the idea that there are only interpretations, interpretations of interpretations, and so on *ad libitum*, with no possibility of introducing objective constraints and differences.

4. THE PRESENT SITUATION

4.1. THE CRISIS OF METAPHYSICS: BEYOND VERIFIABILITY

We can conclude that it looks likely that the recent developments in the philosophy of science have not undermined the great leading ideas of

---

[1] Reichenbach 1958: 35, 37.
[2] Parrini 1983.
[3] Quine 1964d: 41.
[4] See also below, ch II, sec. 5.1.

the positive spirit (relativism, empiricism and anti-metaphysical objectivism), but rather the peculiar linguistic form in which these ideas were formulated in the philosophy of the Circles of Vienna and of Berlin.

I will now turn to the reasons that lead me to think that the abandonment of the standard conception has created a situation which encourages the decision to retain the guiding principles of the positive spirit, thereby resigning ourselves to the dilemma between metaphysical realism and radical relativism. Three salient aspects of the current epistemological scene seem to me particularly important: (1) the absence of dominant philosophical perspectives; (2) the amplification of the effects of the neoempiricist critique of the Kantian theory of the synthetic a priori, including the surrender of all normative and foundational claims at the theoretical and methodological level – a scenario that can be epitomised by Feyerabend's "anything goes"; (3) the persistence of the uneasiness caused by the neopositivist attacks on metaphysical speculation.

As for this last aspect, it is important to recall that, even in its most propagandistic formulations, the neopositivist critique of metaphysics was not a mere accusation of meaninglessness. In the 1929 Manifesto, statements such as "'there is a God', 'the primary basis of the world is the unconscious', 'there is an entelechy which is the leading principle in the living organism'", were declared meaningless. But it was specified that they are so "if one takes them in the way that metaphysicians intend". At the same time, it was recognised that "[o]ne can [...] often re-interpret them as empirical statements; but then they lose the content of feeling which is usually essential to the metaphysician".[1]

This criticism was strengthened by referring to the results of the logical analysis of language and to the conception of logical and mathematical deduction as tautological, and therefore empty, devoid of content. Neurath, Hahn and Carnap maintained that "[i]n metaphysical theory, and even in the very form of the questions, there are two basic logical mistakes".

The first mistake springs from "too narrow a tie to the form of *traditional languages*". As Russell and Wittgenstein's logical-analytic theories show, our habitual modes of expression tend to conceal the logical form of language, thus inducing various kinds of conceptual distortions, erroneous hypostases and category mistakes, all equally pernicious to the development of a linguistically correct philosophy.

The second kind of mistake is connected with the belief in synthetic a priori judgements. It derives from "a confusion about the logical achievement of thought", that is, from the mistaken "notion that *thinking* can either lead to knowledge out of its own resources without using any empirical material, or at least arrive at new contents by an inference from given states of affairs. Logical investigation, however, leads to the result that all thought and inference consists of nothing but a transition from statements to other statements that contain nothing that

[1] Neurath, Hahn, Carnap 1973: 306f

was not already in the former (tautological transformation). It is therefore not possible to develop a metaphysic from 'pure thought'".[1] The 1929 Manifesto suggested that one should look for an explanation of the genesis of the metaphysical urge by means of sociological, psychological and, in particular, psychoanalytic investigation.[2]

A further pillar of the neopositivist critique of purely speculative claims was Schlick's attack on the notion of intuitive knowledge in the *Allgemeine Erkenntnislehre* (1918/1925). After reading Wittgenstein's *Tractatus*, and after the linguistic turn, Schlick reformulated his position in the London lectures of 1932, *Form and Content*, making use of the verifiability principle and of Wittgenstein's ideas on logical form (with which, incidentally, the physicalist Neurath strongly disagreed). But the points I want to underline were already fully developed in the *Allgemeine Erkenntnislehre* and in some of his previous writings. Openly polemicising against the intuitionism of Bergson and of certain aspects of Husserl, Schlick maintained that knowledge concerns only structural relations between things. The intuitive content, instead, is unknowable and inexpressible. The essential nature of knowledge is the recognition of identity in difference by means of the discursive use of an appropriate conceptual framework. Intuition and immediate experience cannot be considered forms of knowledge, because they do not entail the recognition of something as something else. The expression 'intuitive knowledge' (as Russell's expression 'knowledge by acquaintance') is as unacceptable as the expression 'round square': they both connect incompatible notions. Consequently, all attempts to found metaphysics relying on the idea that a cognitive intuition can grasp the ultimate essence of reality, fail.[3]

As Feigl summarised, commemorating Schlick in 'Erkenntnis' in 1938, "[m]etaphysical knowledge is not just impossible because it would overstep the limits of human reason, but because its aim rests upon a misunderstanding of the concept of knowledge. *Intuitive* metaphysics is impossible, because intuition is not knowledge; *deductive* (or dialectical) metaphysics is wrong, since, misunderstanding the nature of deduction, it fancies itself able to arrive at conclusions that are not logically contained in the premisses. *Inductive* metaphysics is not, indeed, logically impossible, so long as the inductive method leads to claims that are basically testable. But if this be so, there is continuity with science, and the danger lies merely in the fact that very risky, fantastic or unfruitful hypotheses may be set up".[4]

In the light of what we have just seen, the crisis of the verifiability principle seems an insufficient reason to ignore the neopositivist attacks on metaphysics, where metaphysics is understood as a discipline

---

[1] Neurath, Hahn, Carnap 1973 307f.

[2] Neurath, Hahn, Carnap 1973: 307

[3] See Schlick 1974, ch. I, 'The nature of knowledge' ['Das wesen der Erkenntnis'] (pp. 1-101).

[4] Feigl 1979: xxif.

that claims to attain knowledge of reality in itself, or of the essential properties of things, by means of a research endowed with some degree of objective, or at least intersubjective, validity.

It is true that the attempts to give an acceptable formulation of the verifiability principle have failed, and this makes it impossible to draw a clear boundary between cognitively meaningful discourse (science) and cognitively non-meaningful discourse (metaphysics). It is also true that Quine's critique of analyticity and meaning has inspired some philosophers to maintain that analytic philosophy is incapable of overcoming or dissolving metaphysical problems: the so-called linguistic analysis allegedly presupposes an interpretation-reconstruction of meanings analogous to the old speculative positions, and as controvertible as them.

But, on close inspection, all this represents a meagre consolation for the metaphysician, who hopes to come up one day with something different from what is already offered us by more or less sparkling essay writing, by the expression of an eminently respectable religious feeling for life, by the communication of mystical intuitions resulting from deep internal concentration, by the autobiographical narration of unresolved personal problems. After the neopositivist attack, metaphysicians can no longer proudly silence those who question the cognitive value of their speculations, by appealing to the fact that we do not possess a rigorous criterion of cognitive meaningfulness. If metaphysicians aspire to be credible, they must specify what is the nature of their intended discourse, showing in which way it differs from the formulation of an exclusively personal emotional point of view, from pure conceptual analysis, from the suggestion of extremely general hypotheses on the nature of being.

This last possibility makes way for the so-called 'influential metaphysics' – an enterprise which, when accomplished, will prove itself to be very close to science, and will be incorporated in it, as Kant's 'Critical metaphysics'. But this is not very much in comparison with the old claims of a "science of being *qua* being". Since they have some content, hypotheses of this kind will be in principle testable even from the empirical point of view, at least conjointly with other components of empirically testable theoretical systems. It is true that these connections with experience, being very indirect and loose, cannot lead to definitive confirmation or disconfirmation; and they do not allow us to draw up a formally definable criterion for the sharp demarcation of science from non-science. Nevertheless, these connections with experience are sufficient to confer strengths of varying degrees on judgements of riskiness, fantasticality, unfruitfulness, etc.. And, for this kind of evaluation, distinctions of degree are more than sufficient.

In conclusion, if we cannot deny the metaphysician's right to hazard fanciful hypotheses, we cannot even deny the anti-metaphysician's right to illustrate the dreams of metaphysics with the dreams of a visionary.

For similar reasons, the present crisis of philosophy as the analysis of the language of science (according to Carnap's most rigid formulations) cannot be interpreted as lending support to the pretension that respectable philosophy can be done without an analytic grounding, without a rigorous analysis of the various forms of culture. Many thinkers are proclaiming the end of the age of analysis. But, in the very rare cases in which this has turned out to be something more than wishful thinking, this has been due to the thorough analytic training of these thinkers. In most other cases, there has only been a relapse into superficial journalistic ideology, conceptually volatile edifying rhetoric, autobiography scattered with references to the history of philosophy.

On the whole, in the present situation, neither of the two rivals seems to be in a particularly good shape. Philosophical analysis has impaired the critique of metaphysics and the concept of philosophy as analysis, in so far as they rely upon the verifiability principle and upon the linguistic theory of the a priori. But this has not fostered purely speculative constructions other than empty verbal exercises and 'narratives' endowed with a depressingly low theoretical content.

4.2. PLURALISM

As well as the effects of the neoempiricist critique of metaphysics, the 'openness' and the 'pluralism' of the present philosophical scene are also relevant to the prospects of positive philosophy.

Almost twenty years have passed since Suppe commented on the end of the standard conception contending that "philosophy of science is coalescing around a new movement or approach which espouses a hard-nosed metaphysical and epistemological realism that focuses much of its attention on 'rationality in the growth of scientific knowledge' and proceeds by the examination of historical and contemporary examples of actual scientific practice".[1] So far, Suppe's prediction has not come true.

The idea of the insufficiency of a purely logical analysis of theories has gained ground: the historical, sociological, psychological and biological dimensions of knowledge must be taken into consideration. (This idea is not completely new, if compared with Neurath's old positions.) The neoempiricist (as well as Popper's) distinction between "context of discovery" and "context of justification", between logic and psychology of research, has been increasingly disputed. All this has coincided with the appearance of many historical and conceptual investigations on the concrete dimensions of 'scientific reason'. These investigations have led progressively farther away from the standard conception of theories, towards an enrichment of our image of science, and in some cases towards the reformulation of various kinds of realism. However, logical positivism, the fundamental reference for analytically

---

[1] Suppe 1977b 618

oriented studies in the philosophy of science from the end of the twenties to the first half of the fifties, has not been followed by the consolidation of a new and different paradigm of analysis.

Contemporary philosophy is on the whole characterised not by a unitary perspective, but rather by a variety of tendencies. Many important specific contributions have appeared, mostly without explicit connections with general perspectives. A proliferation of approaches has developed, inspired by different philosophical positions. Along with those connected to realism, to cognitive sciences and to evolutionary epistemology, there are still to be found empiricist approaches. It is sufficient to think of Mary Hesse's "new emprìcism" (strongly influenced by Quine), Shapere's "contingent empiricism", van Fraassen's "constructive empiricism".[1] The only relatively unitary aspect is a marked attention to rigour of investigation (one of the best legacies of neopositivist epistemology), which characterises especially the studies concerning the dynamics of theories, the nature of scientific explanation and laws, the contrast between inductivism and deductivism, the analysis of specific theories, such as relativity, quantum physics, evolutionary theory and psychoanalysis.

In the present 'conversational' climate, characterised by tolerance and by the absence of hegemonic perspectives, the following tendencies have gained ground: the methodological anarchism of the "anything goes" type; the abandonment of the foundational claims of epistemology; the weakening of the coercive power of philosophical arguments. It would be strange, then, if only the positive philosopher were to be denied the possibility of occupying a credible position. As a matter of fact – as I have pointed out – the crisis of rigid empiricist models, such as inductivism and reductionism, has not resulted in the widespread surrender of the wider empiricist perspective, characterised by the rejection of 'transcendental conditions of the possibility of experience'. On the contrary. This critique of Kant, together with the crisis of the standard conception's logicising formalism, has fostered the recent versions of evolutionary, psychological and sociological 'naturalistic' approaches reminiscent of nineteenth century positivism.

5. POSITIVE VS POSITIVIST PHILOSOPHY

The complex contemporary situation suggests that the crisis of neopositivism should not be made to coincide with the crisis of the fundamental principles of positive philosophy. On the contrary. There are reasons to believe that a marked tendency towards relativism lies at the roots of the recent philosophical conceptions which see positivism as their *bête noir*. The problem is that the stress on relativism, brought about by many (justified) criticisms of the standard conception and the two dogmas of

[1] See, respectively, Hesse 1976, Shapere 1984a, Introduction, in particular pp. xxixf; van Fraassen 1980. 6ff.

empiricism, seems to blur the difference between science and metaphysics, threatening the objectivity of scientific knowledge. Consequently, any reliance on the ideals of the positive spirit will have to deal with these criticisms, by preserving part of their relativist import, whilst emptying them of radically anti-objectivist and anti-empiricist implications. This effort of conceptual mediation will result in what I call 'positive philosophy', as distinct from the old and new positivisms of the past.

Of course, I am aware of the fact that the term 'positivism' derives from 'positive'. I insist on the difference between 'positivism' and 'positive philosophy', because I am convinced that one of the main limits of previous positivisms was to have been not 'positive' enough. That is to say, positivist philosophies were never fully successful in formulating their leading ideas by correct extrapolation from the philosophical analysis of scientific knowledge. It has been acutely noted that a philosophy which keeps only to the data, runs the risk of not grasping the data. I call for a philosophy that is able to grasp the full import of the data, keeping to them as closely as possible, and correctly extrapolating from them its normative ideals.

The elaboration of this position will not be completely painless for some of the typical theses of analytic philosophy of language. It will entail the abandonment of a purely linguistic view of philosophical problems, and the return to a more traditional conception. My move will be analogous to van Fraassen's some years ago with "constructive empiricism". Presenting the ideas of *The Scientific Image*, he maintained "that empiricism is correct, but could not live in the linguistic form the positivists gave it". The logical empiricists "were right to think in some cases that various philosophical perplexities, misconceived as problems in ontology and epistemology, were really at bottom problems about language [...] But this [...] emphatically does *not* mean that philosophical concepts must be one and all linguistically explicated".[1] I think that a positive philosophy is preferable to "constructive empiricism", because it allows one to reject metaphysical realism, without requiring one to give up scientific realism about the non-observable entities of science. But, in a similar way to van Fraassen, I am convinced that positive philosophy cannot survive in the logical-linguistic form it was given by the logical empiricists. Not all the problems of positive philosophy can be seen as questions of logic and of philosophy of language. Philosophical anomalies and riddles are not all born of linguistic misunderstandings. This is true of some of them (Neurath would have said, of many of them); but some philosophical puzzles raise theoretical and conceptual questions that require theoretical and conceptual solutions.

The salient points of my proposal will be two.

(1) I will move, even more decidedly than the previous positivisms (including Viennese neopositivism), towards a conception of truth and of objectivity deeply influenced by Kant. This will not entail

---

[1] Van Fraassen 1980: 3f.

the surrender of the empiricist thesis that all our knowledge is uncertain, partial, and corrigible; on the contrary, I will try to reconcile the best elements from both perspectives.

(2) The connection with Kantianism will allow me to 'ground' my positive philosophy non-metaphysically, so as to shield it from the objection (widely raised against verificationism) of requiring criteria of acceptability so strict as to make itself unacceptable by its own assumptions.

The ladder that I will endeavour to build should be in the end very different from the ingenious, but somewhat overly 'crooked', ladders put together by the 'early' and the 'later' Wittgenstein, which created so many problems first for the Vienna Circle, and then for post-neopositivist philosophy. My aim will be to achieve the maximum balance between the semantic and ontological component and the component connected with epistemology and the philosophy of science. The vertical sides of my ladder must be calibrated so as to achieve an equilibrium as close as possible to that of the ladders we confidently climb up in everyday life.

Chapter II

EPISTEMIC RELATIVISM, APRIORISM, NEOEMPIRICISM

1. EPISTEMIC RELATIVISM AND THE THEORY-LADENNESS OF OBSERVATION

Relativism, the first of the three fundamental ideas of the positive spirit (relativism, empiricism, anti-metaphysical objectivism), seems particularly alive today.

Carnap's relativisation of the criteria of analyticity and cognitive significance makes one suspect that the failures of the logical positivists depended in part upon an initial underestimation of the relativist component of their epistemological position. With Neurath as the only exception, their search for general and formal criteria of scientificity and inductive confirmation saw them for a long time committed to an abstract and metahistorical conception of scientific rationality.

By contrast, in the present philosophical situation the thorniest problems of positive philosophy stem from the strength of the relativist thrust. The discussions of phenomenalist reductionism, observation and scientific methodology have challenged the empirical and objective value of knowledge. This has paved the way for widespread forms of semantic, epistemological and ontological relativism, which seem the only available alternative to metaphysical realism. My concern will be to avoid the excesses of relativism without getting caught up in the difficulties of metaphysical realism.

Relativism is an extremely complex range of doctrines, spanning the domain of theoretical as well as moral philosophy. I will restrict my attention to the problem of knowledge. In Chapter VI I will show that even epistemic relativism can be understood in more than one way, and I will reject its radical version. For the present, I will consider a moderate and readily acceptable notion of epistemic relativism, which I will discuss in connection with the problem of the a priori.

For purposes of presentation, I distinguish between various dimensions of relativism, which I will illustrate by referring to present and past thinkers. I do not imply that these thinkers are relativists *tout court*. To be honest, relativism is not a very attractive general philosophical position: it does not even satisfy completely the majority of those who have brought about the 'relativist side' of the present epistemological scene. I do not intend to classify the authors I will cite: my only purpose is to point out the reasons for epistemic relativity, which I will then try to reconcile with empiricist, objectivist and anti-metaphysical requirements.

Epistemic relativism largely prevailed following the attacks of the new philosophers of science on the standard conception of scientific theories.

The core of epistemic relativism is the idea that scientific knowledge is set within a conceptual scheme more or less uniformly shared by a community of individuals engaged in certain scientific-cognitive practices. This epistemic framework is a research perspective which is closely tied to the structure of the language which is used, incorporates a "thought style"[1] concerning a variety of more or less clearly characterised phenomena, delimits the class of legitimate problems and sets the standards of acceptable solutions. Some of these conditions are "tacit" presuppositions which can be made explicit in the passage from one context of research to another; the new context, in its turn, will be characterised by different tacit factors. There is no reason to believe that there are absolutely fixed and invariant assumptions; however, the change of conceptual schemes requires the more or less explicit adoption of other epistemic assumptions. Certain assumptions constitute the core of the theoretical structure. As such, they represent an epistemologically distinct class of statements, which are neither a priori, since they can be modified according to experience, nor trivially a posteriori, since they are 'protected' from direct empirical refutation.

According to the relativist conception, cognitive, and, especially, scientific, activity "is done from within a *Weltanschauung* or *Lebenswelt* and the job of philosophy of science is to analyze"[2] the linguistic-conceptual, theoretical and methodological-axiological systems within which science operates. This approach is connected with all the philosophical tendencies which, following Kant, have emphasised the subjective components of cognitive activity. However, the connection with the pragmatist tradition and historicism sets relativism apart from Kant: the uniqueness of the categorical system is denied, and great importance is attributed to the historical change of categories; special attention is payed to the sociological factors which influence the acceptance, the formulation and the rejection of world-views.

It is significant that relativism has established itself by means of a polemic – albeit not always to the point or justified – with the standard conception, which was accused of presupposing the theoretical neutrality of the observational basis, of artificially separating the discovery of hypotheses from their justification, of limiting itself to the logical analysis of the structure of theories, neglecting the dynamic aspects of knowledge.

My aim is only to underline the dependence of the cognitive enterprise upon a more or less organic complex of different subjective epistemic conditions. The plausibility of this thesis depends upon the existence of a marked gap between experience and the complex of our "beliefs and convictions". A number of arguments have been deployed in order to establish this separation. I will concentrate on the one which proves the theoretical character of observation by disputing the di-

---

[1] Fleck 1979.

[2] Suppe 1977a: 126.

chotomy between observational and theoretical language, which Carnap laid at the basis of the standard conception.

1.1. CRITIQUE OF THE 'SEMANTICAL CONCEPTION OF OBSERVATION'

Many arguments have been used to prove the thesis of the theory-dependence of observation. Some have questioned the supposed 'incorrigibility' of basic statements, others have looked at the functioning of language, the distinction between observable and non observable entities, the conception of perception and belief, the great 'revolutionary' scientific changes. Most of these arguments are characterised by an intricate mixture of all these closely connected perspectives. Setting aside the objections of Achinstein, Maxwell, Putnam, Hanson, Toulmin and Kuhn,[1] I will concentrate on the critique developed by Feyerabend starting with the essay of 1957-58, 'An Attempt at a Realistic Interpretation of Experience'. My discussion of this critique is meant to highlight the fundamental philosophical aspect of the matter, to which I will refer in the next chapter in order to secure the empirical value of knowledge. The attentive reader will see in Feyerabend's arguments something more than the mere echo of Husserl's, Cassirer's, Wittgenstein's and Popper's criticisms of empiricist theories of abstraction.

The main target of Feyerabend's criticism is the 'semantical conception of observation'. The semantical conception of observation is, in my view, an abstract theoretical position which cannot be identified *tout court* with logical empiricism, despite the fact that it was incorporated in the more rigid formulations of the verifiability principle. The semantical conception of observation says that empirical observation is the primary source of meaningful contents. Experience supplies us with the tool with which to establish directly the meaning (and therefore the truth value) of the so-called observational expressions (for example, 'Red'), and, more or less indirectly, the meaning of the so-called theoretical expressions (for example, 'electron').

In order to avoid old misunderstandings, we must stress that Feyerabend, in criticising the semantical conception, was not denying the existence of an indeterminate, exclusively pragmatic, distinction between theory and observation. Significantly, the starting point of his argument is a 'pragmatic theory of observation', whose purpose it is to specify the "conditions which a language must satisfy in order to be acceptable" as an 'observation language', i.e., "as a means of describing the results of observation and experiment". This theory is in certain respects a development of the ideas upheld by Neurath in the controversy over protocols. It is not to be confused with the 'principle of pragmatic meaning', which I will talk about briefly, nor with the 'pragmatic con-

[1] See the detailed reconstruction of the debate in Suppe 1977a; 1977b; 1989, p. I, 'From the Received View to the Semantic Conception'.

ception of observation', which I will illustrate in the next chapter, incorporating it in my own proposal.

At the roots of the *pragmatic theory of observation*, there is the distinction between "two sets of conditions for observation languages": (1) the 'characteristic' and (2) the 'interpretation'.

1. The 'characteristic' – "The conditions of the first set are *pragmatic* (psychological, sociological) conditions. They stipulate what is to be the relation between the (verbal or sensory) behaviour of human beings of a class *C* (the observers) and a set of physical situations *S* (the situations observed)". Given three classes, *A*, *C* and *S*, *A* will stand for the class of observable (atomic) sentences used by the class of observers *C* in the class of situations *S* if, and only if, given an observable situation *S*, every observer *C* "is able to come to a quick, unanimous and relevant decision with respect to those *A* for which the chosen *S* is appropriate." Therefore, "[t]he characteristic of an observation language completely determines the 'use' of each of its atomic sentences".[1]

As Suppe pointed out, from this characterisation it follows that:

(a) "[t]he observation language for a class *C* of observers will be, roughly, those sentences which members of *C* are *caused* to accept or reject in response to sensory phenomena, where there is a consensus among the members of *C* whether to accept or reject the sentence";

(b) "[s]uch assent or dissent is to *uninterpreted sentences*, not statements". Therefore, "such assent or dissent leaves open the question of the interpretation of sentences in the observation language and" it is possible that different members of *C*, though converging in their manifestations of assent or dissent, may assign "different interpretations to the same sentences".[2] (All this is clearly connected to Quine's conceptions of stimulus-meaning and the systematic indeterminacy of translation).

2. The 'interpretation' – In Feyerabend's view, in order to have a language in a full sense, interpretation must be added to characteristic. The pragmatic conditions summed up in the characteristic "concern the relation between observation *sentences* (*not* statements) and human beings without making any stipulation as to what those sentences are supposed to assert". In order to establish the assertive content, it is necessary to add to the pragmatic conditions interpretative conditions which allow us to talk about what the sentences mean. Therefore, "[a] particular observation language is completely specified by its characteristic together with its interpretation".[3]

Feyerabend's thesis is that the *characteristic* alone does not determine the *interpretation,* and that the interpretation, in its turn, depends upon assumptions of a theoretical nature. This conclusion is justified by showing the untenability of the semantical conception of observation. To be more precise, Feyerabend criticises the two principles

[1] Feyerabend 1981b: 17f.

[2] Suppe 1977b: 637.

[3] Feyerabend 1981b: 18f.

that would make it possible to state that the meaning of observation sentences (and their truth value) is *logically* determined by observation, by the conditions of their empirical application. The principles in question are those of (1) 'pragmatic' and (2) 'phenomenological meaning'.

1. Critique of the principle of pragmatic meaning – The refutation of the principle of pragmatic meaning is already contained in the distinction between characteristic and interpretation. According to the principle of pragmatic meaning, the meaning of linguistic expressions is determined by the circumstances of their use. In this view, the interpretation of the sentences of the observation language is determined univocally and completely by the characteristic, that is, by the set of the conditions of their empirical application. This cannot be for two simple reasons:

(a) "[T]he existence of a certain observational ability [...] is compatible with the most diverse interpretations of the things observed":[1] "[w]hat the observational situation determines (causally) is the acceptance or the rejection of a sentence, i.e. a physical event".[2] But – as will result more clearly from the criticism of the principle of phenomenological meaning – it is impossible to infer from this that the acceptance or the rejection of the sentence is *logically* determined by the observational situation.

(b) Because of the well known problem of induction, "no set of observations is ever sufficient for us to infer (logically)" an interpretation: "however well behaved and useful a human observer may be, the fact that in certain situations he (consistently) produces a certain noise, does not allow us to infer what this noise means".[3] No matter how many observations we collect, it will never be possible to infer (*logically*) the interpretations.

2. Critique of the principle of phenomenological meaning – According to the principle of phenomenological meaning, "interpretations are determined by what is experienced": "the meaning of an observational term is determined by what is 'immediately given' at the moment of the acceptance of any observational sentence containing that term". In conformity with the theory of ostensive definition, in order to explain to someone the meaning of 'red', it is only necessary to bring about the situation in which one has an immediate perception of red.

Three kinds of objections can be raised against the principle of phenomenological meaning, reflecting 'rationalist' criticisms of radically empiricist theories of the formation of concepts.

(a) By means of a trivial application of the 'third man argument', it is possible to show immediately that, for an observer C to be able to consider an immediately given situation (or phenomenon) S as determining the meaning of a declarative expression A (for example, of

---

[1] Feyerabend 1981b: 22.
[2] Feyerabend 1981b: 19.
[3] Feyerabend 1981b: 22.

'red' in the presence of red), a "relation of phenomenological adequacy" $R(S,A)$ must hold between the phenomenon and the sentence. But, at the moment in which A is uttered, the relation of phenomenological adequacy "cannot be immediately given in the same sense in which" S "is immediately given, i.e. it cannot be a phenomenon": the "relation of phenomenological adequacy is not part of the experience of" C.

Therefore, "the utterance of a certain observation sentence cannot be justified by saying that it 'fits' the phenomena. For if by appealing to the relation of phenomenological adequacy we make it part of our experience, we have thereby changed the original phenomenon. And our description of the new phenomenon will still be in need of justification". Thus we become caught in an infinite regress. Of course, it is true that, as observers, we have experience of S. But to repeat this (as apparently Feigl used to do) is of no avail, "for the question discussed is not what is experienced, but whether what is experienced has been described adequately".[1]

(b) It is certainly true that sentences which stand in a relation of phenomenological adequacy with the phenomena have an interpretation. But, unless we put the cart before the horse, we must acknowledge that interpretation "is not conferred upon" linguistic signs "because they 'fit' [the phenomena], but it is an essential presupposition of the 'fitting'". We can say that 'red' is a name which befits red things as distinct from things of another kind only because we already know the meaning of the word 'red'.

(c) No set of observations will ever be sufficient, on its own, to logically infer a particular interpretation. This depends especially upon two reasons:

(i) The existence of so-called 'secondary interpretations': "I may feel a strong inclination to call the vowel *e* 'yellow'. The important thing is now that I feel this inclination only if 'yellow' carries its usual meaning. But according to this usual meaning 'yellow' is not applicable to sounds".

(ii) Given a phenomenon, it is possible to describe it in infinite ways, without the principle of phenomenological meaning being able to pick out the observationally correct description (note that, "if we consider signs in isolation, then any interpretation which we confer upon them is a matter of convention").[2]

Considered on their own, phenomena do not have the 'normative' power to 'determine' the correct interpretations, to 'legislate' on meanings. Putnam too has come to see this, abandoning the causal theory of reference.[3] The same situation can be described in many different ways, according to how we use words: there are no descriptions endowed *per se* with a 'normative value' qua they supposedly

---

[1] Feyerabend 1981b: 25f.
[2] Feyerabend 1981b: 26f.
[3] See, for example, Putnam 1988: 114

describe the essence of things and of their relations. For Putnam, after the conversion from 'external' to 'internal realism', experience is not normative even in the case of extremely general notions, such as object, existence, logical constant.

It remains trivially true that strict adherence to an interpretation and rejection of all different accounts may lead to a psychological situation where the relation between phenomena and meanings is felt as univocal. Nevertheless, there is the logical possibility of alternative interpretations; and the choice between rival interpretations will have to be made on the basis of criteria which are external to the phenomena considered *per se*.

We must conclude that the observed phenomena can be considered at best a (phenomenological) cause of the acceptance or the rejection of a certain sentence *A*. But neither the conditions of use of observation sentences, nor the phenomena accompanying their application in observational situations, can justify logically the interpretation of such sentences and the truth value of the corresponding statements.

2. THREE DIMENSIONS OF EPISTEMIC RELATIVISM

What has been said in the previous section raises the problem of the ways of the specification of interpretation, and of the relations holding between constitution of meaning, theoretical elaboration and empirical observation. I will discuss the matter in the following chapter, opting for an answer which is compatible with an empiricist perspective.

For the present, I will consider the relativist implications of the divide between experience and our linguistic descriptions. I will illustrate the reasons why it is not possible to infer the complex of our statements about reality from the empirical data without referring to a set of linguistic, theoretical and methodological-axiological epistemic conditions, entailing corresponding forms of relativism.

2.1. LINGUISTIC RELATIVISM

In Kantian terms, 'linguistic relativism' is the ascription of a sort of constitutive function to languages. The attribution of an epistemic role to language means that language is a conditioning factor in the cognitive process, a medium that moulds the way we experience the world. Languages are not simple vehicles of communication, made up of a set of phonetic and graphic 'labels' used to indicate ready made entities and cognitive contents. Language cuts out, or carves, experience in ways which are not univocally determined by the intrinsic qualities of our sensations or by the hypothetical joints of reality in itself. Language is a structure that contributes to organising the empirical material in a more or less well ordered system of objects in relation with each other.

Cassirer rightly valued Humboldt's powerful metaphor: "By the same process whereby [man] spins language out of his own being, he ensnares himself in it; and each language draws a magic circle round the people to which it belongs, a circle from which there is no escape save by stepping out of it into another".[1] "The difference between the several languages [...] is not a matter of different sounds and marks, but of different world conceptions".[2] And the linguistic differences in the representation of reality are closely connected with differences in the actions aimed at the satisfaction of practical-vital interests. In a way analogous to the later Wittgenstein's connection of linguistic uses and forms of life, Cassirer maintains that "the recognition of function precedes that of Being. The aspects of Being are distinguished and co-ordinated according to a measure supplied by action – hence they are guided, not by any 'objective' similarity among things, but by their appearance through the medium of practice, which relates them within a purposive nexus".[3]

The extreme relativist consequences of this conception of language have been spelt out by linguists such as Whorf and Sapir. Sapir's position is multifaceted. Willing to deny the existence of a connection between race and language, he affirmed that "[t]he latent content of all languages is the same – the intuitive *science* of experience". At the same time, he insisted on the unseverable tie between the language, our "thought-grooves" and culture (for example, "[a] society that has no knowledge of theosophy need have no name for it").[4] Sapir came to the radically relativist persuasion that the "worlds in which different societies live are distinct worlds, not merely the same world with different labels attached"[5] – a statement that matches Kuhn's "exclamation" "that when paradigms change, the world itself changes with them".[6]

To put it another way, *deep linguistic differences are connected with different realities, not with different ways of relating to the same reality*. Languages do not play a constitutive role with respect to the *ways* in which we experience the world, but they entail a plurality of different worlds. As Whorf puts it, whereas according to the common conception "'Sentences are unlike because they tell about unlike facts'", according to "the point of view of linguistic relativity, [...] 'Facts are unlike to speakers whose language background provides for unlike formulation of them'".[7]

Recently, some anthropologists, sociologists and philosophers of language have concluded from Wittgenstein's analysis of 'following a

[1] W. von Humboldt, *Einleitung zum Kawi-Werk,* S.W. (Coll. ed.), VI, 60, cited in Cassirer 1953c 9.
[2] Cassirer 1953c. 31.
[3] Cassirer 1953c: 39.
[4] Sapir 1982. 217-9.
[5] Sapir 1951: 162.
[6] Kuhn 1996a: 111.
[7] Whorf 1964: 235.

rule' that the only reality we know is the reality we see through the coloured spectacles of our language. People who speak different languages live in different realities and know different truths. Each society has its particular vision of reality, its specific universe of thought, and, ultimately, its universe of truth. Reality is 'socially constructed'. The science created by Western civilisation has produced another language in addition to the pre-existing ones, and the scientific reality and the truths resulting from it are such only for those who agree to speak that language.

Quine too, in *Word and Object*, referred to Cassirer, Sapir and Wittgenstein in order to argue "that deep differences of language carry with them ultimate differences in the way one thinks, or looks upon the world".[1] However, as we will see, he does not really compromise himself with the relativisation of reality and truth as such. What he does is to ingeniously connect linguistic relativism with the principle of systematic indeterminacy of translation, in order to expose the limits of a certain semantical Platonism and mentalism which conceives of meanings as entities of which words are the labels. He criticises, that is, the "illusion that our so readily intertranslatable sentences", belonging to the languages of Western culture, "are diverse verbal embodiments of some intercultural proposition or meaning, when they are better seen as the merest variants of one and the same intracultural verbalism". In order to realise the inadequacy of all conceptions which see meanings as entities independent of the different intracultural verbalisms they are associated with, it is sufficient to reflect about what would happen if we tried to translate a sentence such as 'Neutrinos lack mass' into the language of a tribe whose dominant activity was constituted by elementary forms of agriculture: there is no "free-floating, linguistically neutral meaning which we capture, in 'Neutrinos lack mass', and the native cannot".[2]

Analogous theses have been upheld in philosophy of science, for example, by Hanson. Physics is characterised by a 'locking' of concept and language entailing a dependence of the capacity to see or grasp certain facts upon the apparatus of concepts and meanings which is used. The fact that different people make use of different languages entails differences in the things they see and the facts they believe: "[p]erhaps facts are somehow moulded by the logical forms of the fact-stating language. Perhaps these provide a 'mould' in terms of which the world coagulates for us in definite ways".[3]

2.2 THEORETICAL RELATIVISM

A. Language presupposes theory

In philosophy of science the most interesting argument for the moulding function of languages with regard to experience, entails the scrutiny

---

[1] Quine 1970: 77.
[2] Quine 1970: 76.
[3] Hanson 1965: 36.

of the relations between language and theory. It has become clear that the formation and use of a linguistic system presuppose the construction of a theory and implicit reference to theoretical assumptions.

Some thinkers of the past (Campbell, for example) professed this thesis in forms which suggested an unwarranted connection between the attributions of meaning and the acceptance of the truth of certain beliefs: "[w]e [...] invent new terms (or give a new meaning to words previously in common use) when we have new facts or ideas to express; these new facts are usually laws or propositions which again depend for their meaning on laws, and the terms are useless and meaningless unless the laws are true. Thus, if we say anything about electrical resistance we assume that Ohm's Law is true; bodies for which Ohm's Law is not true, gases, for example, have no electrical resistance".[1] Setting aside the connection between truth and meaning, the idea advocated by Feyerabend is right: languages, both scientific and common, have been introduced in order to express some theory or point of view. Therefore, they contain a well-developed and sometimes very abstract ontology.[2]

This thesis had already been illustrated very clearly by Duhem in debate with Poincaré. It is not possible to construct the language in which physical theories are formulated without the previous construction of these very theories, intended as structured sets of theoretical principles giving a nomological value to the observational generalisations which can be made at the level of pre-theoretical experience. It is because we have been able to discover some empirical connections concerning electrical phenomena, and to interconnect them by means of theoretical elaboration, that it has been possible for us to develop the language in which the theory of electricity is commonly formulated. It is not a coincidence that scientific terms have no meaning for those who do not know the theoretical constructions they belong to, and the many ways in which scientific terms relate to the phenomena constituting their domain of empirical application. In Hempel's terms, for Duhem the formation of concepts, or meanings, is not possible without the formation of theories.

I will risk an analogy with one of the central theses of Kant's *Critique of Pure Reason*. In § 15 of the 'Transcendental Analytic', Kant says that the act of synthetic combination (*conjunctio*) of the manifold of representations "is an act of the self-activity of the subject", which cannot be given by objects, and the dissolution, or "*analysis*, which appears to be its opposite, [...] always presupposes it. For where the understanding has not previously combined, it cannot dissolve, since only as having been combined *by the understanding* can anything that allows of analysis be given to the faculty of representation".[3] Similarly, for Duhem the construction of a theoretical language, and of a complex of purely linguistic (analytic) relations between the terms of the vocabu-

[1] Campbell 1957: 43
[2] Putnam 1975c: 117-22, 124-6, 128-31.
[3] Kant 1985: 151f [B 129f].

lary, presupposes an autonomous intellectual elaboration. This elaboration, starting from experience, constructs a theory which encompasses the manifold of empirical phenomena under an abstract and symbolic unity.[1] This analogy is supported by the fact that reference to Duhem's epistemology is one of the pivots of Cassirer's criticism of Mach's conception of knowledge (interpreted as a radical form of sensationalist empiricism).[2]

B. Theoretical assumptions

Contemporary philosophy of science emphasises the role of the linguistic framework as well as the dependence of cognitive activity upon explicitly theoretical presuppositions, which sometimes are conventional (but not linguistic) assumptions. This kind of 'theoretical relativism' has emerged clearly from the analysis of the manifold aspects of the relation between experience and theory: the logical structure of empirical testing, the justification of induction, the status of the protocol statements, the role of idealisation, the presuppositions of the method of relevant variables, of the repeatability of experiments and of eliminative induction, the use of *ceteris paribus* and conditional clauses.

Using freely Goodman's words, we can epitomise theoretical relativism by saying that our judgements of the empirical adequacy of our sentences, of their validity, of their degree of confirmation, of their truth value, do not depend only "upon what is presented" by experience, but also "upon how it is organised", both by means of the language we use and by means of a set of theoretical presuppositions. These presuppositions, far from being inexorably imposed by the "nature of human cognition", are hypothetical assumptions open to criticism and revision.[3]

As Hempel has recently shown on the basis of a clear analysis of conditional clauses, the choice and evaluation of assertions "[o]n what there is, or what the world is like" are tied to the background knowledge constituted by "the best world view, or the best theoretical system, we have so far been able to devise".[4]

The best introduction to theoretical relativism is represented by the traditional problems of inductive confirmation and of the logical structure of empirical testing. Laudan has called them, respectively, (1) the 'plebeian' and (2) the 'aristocratic' problem of induction.[5]

1. The 'plebeian' problem of induction – The 'plebeian' problem of induction concerns the possibility of justifying 'direct' empirical generalisations, that is, the problem of inductive inference from known

---

[1] See Duhem 1962a, p II, ch IV, sec. 2, 'The result of an experiment in physics is an abstract and symbolic judgment' (pp. 147-53).

[2] Cassirer 1950, p. I, ch. V, 'The Goal and Methods of Theoretical Physics' (pp 81-117)

[3] Goodman 1983. 96f.

[4] Hempel 1988: 14.

[5] See Laudan 1981· 73f.

past cases to future unknown ones. After Hume's classical formulation, the issue has acquired new dimensions with the discovery of Hempel's and Goodman's paradoxes and with the formulation of the connections between probability and inductive confirmation. I will not examine in detail the issue and the various ways, 'Pascalian' or 'Baconian', in which it has been treated.

The point I want to make is very general, and it is solidly established in the vast recent literature on induction and probability, whatever the approach that is followed (including Popper's deductivism, which aspires to replace the notion of inductive confirmation with that of corroboration). Since Hume's time, there have been increasingly sophisticated formulations of the problem, fully exploiting the mathematical theory of probability. However, from the general philosophical point of view no real progress has been made since Hume. *Inference and inductive confirmation are possible only on the basis of theoretical and / or methodological background assumptions*: this conclusion is invariably the same, whether we use Mill's eliminative induction of relevant variables, or Bayes' theorem, or whether we aim (as in a recent contribution by Earman) towards a "Bayesian eliminativism" founded on a sort of "partnership between Holmes and Bayes".[1]

As I will explain in the final chapter, I do not mean to suggest by these comments that the great amount of work which has been done is not instructive from a general philosophical point of view. On the contrary, what I want to say is that this work has not affected the idea that inductive practice (as well as Popper's deductivism) requires the use of presuppositions of various kinds. These presuppositions include: the prescription of appropriate limitations, gauged on the background knowledge, to inductively projectable predicates; the acceptance of methodological rules concerning the goodness of hypotheses; the preventive attribution of degrees of probability or plausibility to the various conjectures on the relevant variables; the reference to theoretical assumptions ('metaphysical' according to some) about supposed structural characteristics of reality, such as the uniformity of nature, the way in which the world is divided into a limited variety of species, and so on. In conclusion, the evaluation of inductive generalisations depends upon theoretical or methodological assumptions which, in their turn, cannot be further justified, under pain of an infinite regress like the one exposed by Hume.

It has been widely shown that this holds both for universal and probabilistic empirical generalisations. Accordingly, it is not possible to bypass the problem by considering the confirming cases as a proof of the probability, rather than a proof of the truth, of the conclusion of the empirical-inductive inference. This is extremely clear from the many works written on the subject by the major theoreticians of the recent past, from Reichenbach to Keynes, from De Finetti to Carnap, as well as

[1] Earman 1992, in particular pp. 163, 170, 183-5.

by more recent thinkers such as Jeffreys, Shimony, Burks, Salmon, Hesse, Hintikka, Kyburg, Jeffrey, L.J. Cohen, Levi, Mackie, etc.

2. The 'aristocratic' problem of induction and the holistic character of empirical testing – This theoretical relativism can be further clarified by looking at the 'aristocratic' problem of the empirical testing of our hypotheses and theories. While in 'plebeian' induction favourable cases are *direct* proofs of the validity of empirical generalisations, in the evaluation of theoretical statements, favourable outcomes are *indirect* proofs of their empirical adequacy.

As Hertz saw, in the case of abstract symbolic representations, what is commonly done is to test the predictions about observable events which can be logically inferred from the theoretical statements.[1] To use one of Laudan's examples, the verification of the "statement that the doubling of the volume of a gas halved its pressure, is a positive instance of Boyle's law", according to which the relation $pV = k$ holds; but "it is (in a technical sense) *not* a positive instance of" the assumption with which Newton tried to explain Boyle's law, namely, that gases are made up of "imperceptible particles which repel one another with a force inversely related to the distance between them". That is, such verification is only an indirect proof of Newton's hypothesis: it is not a direct proof in the way that a black raven is a direct proof of the inductive generalisation 'All ravens are black'.[2]

This characteristic of empirical testing raises problems of two kinds.

(a) The first issues from the fact that on the basis of deductive logic it is not possible to infer the truth of a premise (in our case, of an hypothesis or a theory) from the truth of its consequences, since other premises compatible with the same consequences are always logically possible. Therefore, in the case of successful tests it is possible to talk about confirmation of our hypotheses and theories only in the light of assumptions – similar to those mentioned for the justification of induction – which allow us to reject possible alternative explanations.

(b) But in the aristocratic case, theoretical relativism takes on a further, very important aspect, connected with Duhem's rejection of crucial experiments. In Duhem's holistic conception, the testing of a theoretical hypothesis requires a whole series of statements, which are needed in order to deduce a directly testable statement from the hypothesis, to carry out the experiment and to interpret its result. Given this interconnection between the various components of a theoretical system, and given the relative symbolic autonomy of these components from the complex of pretheoretical empirically ascertainable "practical or brute facts", "[t]he only experimental check on a physical theory which is not illogical consists in comparing the *entire system of the physical theory with the whole group of experimental laws*, and in judging whether the latter is represented by the former in a satisfactory

[1] Hertz 1956: 1f.
[2] Laudan 1981: 73f.

manner".[1] When we come across a phenomenon which conflicts with the theory, we will be able to infer by means of *modus tollens* the falsity of the theoretical system as a whole, but not of this or that isolated hypothesis.

Of course, Duhem was aware that the physicist will ascribe the falsification to a specific theoretical component rather than another. Moreover, he did not intend to contest the legitimacy of such a procedure, which is a rather obvious aspect of scientific investigation. The point that Duhem was interested in making was the dependence of the physicist's conclusion upon the implicit admission of the "accuracy of all the other propositions he has used"; "the validity of his conclusion is as great as the validity of his confidence" in these propositions.[2] It is not possible to avoid this relativism by saying that the statements implicitly assumed as true in one context have already been, or will be, tested in another: if Duhem's holism is correct, it is clear that even further empirical proofs will depend upon some auxiliary assumptions.

Of course, the scientist, faced with a certain experimental result, may have no other choice than to ascribe the falsification to one component rather than to another, because the scientist is not able to imagine alternatives of any sort, even of the most implausible kind. (Think, for instance, of the impossibility of starting from non-Euclidean presuppositions when non-Euclidean geometries were not yet available). But this does not prove the falsity of the component that the scientist is forced to abandon because of the lack of alternatives conceivable at the time. This proves only that on some occasions theoretical principles can force themselves upon us with a sort of subjective 'mental necessity', not to be confused with the objective necessity which Kant attributed to the synthetic a priori principles and tried to establish by means of the transcendental deduction.

Therefore, Duhem's conception lends support to the general idea at the basis of theoretical relativism. We are unable to individuate "some unique or direct relation between states of affairs and those hypotheses for which they are taken as evidence [...] What determines whether or not someone will take some fact or alleged fact, *x*, as evidence for some hypothesis, *h*, is not any natural (e.g. causal) relation between the state of affairs, *x*, and that described by *h*, but are other beliefs that person has concerning the evidential connection between *x* and *h*. [...] To put it another way, states of affairs are taken as evidence in light of regularities discovered, believed, or assumed to hold. The evidential relations into which a given state of affairs can enter will thus be as varied as the beliefs about its relations with other states, or as the beliefs about the connections between a class of states to which it can be said to belong and other states (or classes of states)".[3]

---

[1] Duhem 1962a: 200; see also 134f, 149f.
[2] Duhem 1962a: 185.
[3] Longino 1979. 37.

Only reference to theoretical, as well as linguistic, epistemic conditions allows us to overcome the systematic ambiguity of the available data of experience. It seems to me that on the whole this idea has not been affected by the many criticisms raised against Duhem's thesis, which was made famous by Quine. For instance, Duhem's thesis is incorporated from the beginning into Glymour's 'bootstrap strategy'[1] which helps us understand how, in concrete research contexts, one seeks empirical proofs which can be considered as confirmations or disconfirmations of a specific hypothesis with regard to some theory.

2.3. METHODOLOGICAL AND AXIOLOGICAL RELATIVISM

Theoretical relativism is connected with the last level of epistemic relativism, methodological and axiological relativism. It is possible to pass from one to the other by further specifying Duhem's often misunderstood position. In formulating holism, Duhem did not intend to dispute or lessen the importance of experimental results (crucial or not) for the development of scientific research. Considerations of two kinds will make this clear.

(1) One must not forget that Duhem admitted the possibility of falsifying experiments in the case of empirically self-sufficient global theoretical systems. Duhem only denied: (a) that the falsification of a theory could be considered the conclusive verification of a rival theory, because of the impossibility of proving that the latter theory was also the only logically conceivable alternative; (b) that from the falsification of a complex of interconnected statements it was possible to identify univocally the part responsible for the falsification: this identification is relative to some components being assumed as contextually valid.[2]

Incidentally, this makes it very difficult to consider a recent rigorous analysis of the discovery of parity nonconservation in the 1950s as a refutation of Duhem's thesis. What is really done in this analysis is that some logically possible theoretical alternatives are rejected as the product of an irrelevant sceptical attitude, the validity of the relevant observational findings is admitted, and the reference to theories confirmed by empirical proofs independent of the hypothesis which is being tested is considered non-problematic.[3]

(2) In illustrating the holistic conception, Duhem proved his awareness of the fact that in the history of science one often hears of experiments believed to be crucial, that is, of experiments whose outcome – to say it with Hempel – exerted "a decisive influence upon the

---

[1] See Glymour 1980; see also Earman's Bayesian appraisal in Earman 1992: 68, 73ff.

[2] See Duhem 1962a, p. II, ch. VI, sec. 2, 'An experiment in physics can never condemn an isolated hypothesis, but only a whole theoretical group' (pp. 183-8), and sec. 3, 'A "crucial experiment" is impossible in physics' (pp. 188-90).

[3] Franklin 1986.

direction of subsequent theorizing and experimentation".[1] So little did Duhem want to deny this obvious fact that he devoted a lot of attention to the reasons which preside over scientists' choices.

On this subject, Duhem refers to a complex of rules constituting the 'scientific method' and guiding empirical and theoretical research. For Duhem, the provisional and corrigible choice of hypotheses depends upon the acceptance, suggested by *bon sens*, of those values which are commonplace in all the works on philosophy of science and scientific methodology.[2] These values are well known, although not univocally characterisable in a general form: simplicity, fertility, systematic compactness, predictive power, and so on. Their acceptance, in its turn, yields methodological rules which prescribe preference for simpler systems, avoidance of *ad hoc* hypotheses, appreciation for theories which foster new and unforseen empirical predictions, and so on.

The application of these rules can, in its turn, lead us to consider an experimental result as a confirmation of the simpler theory and of the hypotheses which make it up. This presupposes the descriptive assumption that reality, or nature, is simple in the assumed sense of simplicity – in the same way as the positive cases of a universal or stochastic generalisation prove the validity of the generalisation only on the basis of the theoretical assumption, let's say, of the uniformity of nature.

It should be clear that, by subscribing to Duhem's thesis, I do not intend to deny that in certain experimental circumstances some logically possible alternatives may appear to scientists as not very plausible, extravagant, even ludicrous, or such as to imply the endorsement of scientifically irrelevant forms of philosophical scepticism. I am interested only in the relativist implications of Duhem's thesis: I am convinced that these implications raise a philosophical problem that requires a philosophical answer.

However, before I pass on to the reconciliation of the reasons for relativism with empiricism and objectivism, I must say something about three clusters of fundamental questions:

(1) attempts à la Davidson to neutralise relativism (section 3, below);

(2) the legitimacy and the sense of the distinction between linguistic and theoretical relativity, despite Quine's criticism of intentional notions (sections 4 and 4.1., below);

(3) the connections between relativism, the naturalisation of epistemology, and the problems of the a priori and of the justification of our cognitive claims (section 4.1., below).

---

[1] Hempel 1966: 28.

[2] See Duhem 1962a, p. II, ch VI, sec. 10, 'Good sense is the judge of hypotheses which ought to be abandoned' (pp. 216-8), and ch. VII, 'The Choice of Hypotheses' (pp. 219-70).

3. THE POSSIBILITY OF ALTERNATIVE CONCEPTUAL SCHEMES

Attempts à la Davidson to neutralise relativism try to undermine the idea of alternative conceptual schemes upon which the relativist thesis rests. This is achieved by (1) assimilating the notion of scheme to that of language and (2) by defending two statements which should undermine relativism. These statements are: (a) if two or more languages are intertranslatable, they cannot be considered alternative; (b) if it is not possible to translate a complex of sounds and marks into our language, we are not entitled to think that we are faced by a complex of signs constituting a language. Accordingly, it is not possible to have alternative conceptual schemes, because either we have translatability into our own language, or what we are dealing with cannot be considered a language. This does not entail that "all speakers of language [...] share a common scheme and ontology. For if we cannot intelligibly say that schemes are different, neither can we intelligibly say that they are one".[1]

This line of argument, well formulated and made famous by Davidson, has the merit of exposing a rather disconcerting aspect of relativist conceptions à la Whorf, where relativism is argued for by showing, with the tools offered by the Indo-European languages, how different our system of thought is from that inherent in the languages of native American tribes. Despite this, Davidson's argument is not very convincing for at least two reasons.

(1) In the first place, I cannot convince myself (and I am not the only one) that the idea of a language which is untranslatable into our own is logically or conceptually contradictory. I believe it is possible to consider linguistic even vocalisations that we are not able to interpret: after the failure of many attempts of translation, we could be induced to believe, not that the members of a certain community do not communicate linguistically among themselves, but that they do it by means of a language – and therefore with a conceptual scheme – into which we are not able to 'catapult ourselves' using our own concepts as a fulcrum (the image is Quine's).[2]

(2) Secondly, I do not see how it is possible to entirely reduce the idea of conceptual scheme to the idea of language. To do so means to remain caught in the 'panlinguism' of the linguistic turn of philosophy. This entails neglect of a fundamental aspect of epistemology, namely, that the notion of conceptual scheme includes something more than reference to a symbolic apparatus. Cognitive activity occurs within a framework including also synthetic presuppositions, methodological rules and values. And it is obvious that there can be differences in the set of theoretical statements, values and rules assumed as guiding principles, as well as in the way of 'weighing' them in the application to concrete cases. Accordingly, differences of conceptual schemes are possible even within the same languages, whether or not they are reciprocally

[1] Davidson 1984: 198.
[2] Quine 1970: 70.

translatable (it is sufficient to think of the philosophical problems concerning the possibility of using alternative logics and geometries in contemporary physics).

4. EPISTEMIC JUSTIFICATION, CONTEXTUAL APRIORISM, AND THE PROBLEMS OF NATURALISM

4.1. AGAINST QUINE'S REFUSAL OF ANALYTICITY AND INTENTIONALITY

In order to understand the second question (the legitimacy and the sense of the distinction between linguistic and theoretical relativity), it is useful to recall the content of epistemic relativism, bearing in mind the conclusions of Quine's critique of the reductionist version of the verifiability principle.

At least in some formulations dating from the fifties, Quine maintained that the two dogmas of analyticity and reductionism are "at root identical".[1] According to 'Two Dogmas of Empiricism', "[i]t is obvious that truth in general depends on both language and extralinguistic fact. The statement 'Brutus killed Caesar' would be false if the world had been different in certain ways, but it would also be false if the word 'killed' happened rather to have the sense of 'begat'. Thus one is tempted to suppose in general that the truth of a statement is somehow analyzable into a linguistic component and a factual component. Given this supposition, it next seems reasonable that in some statements the factual component should be null; and these are the analytic statements",[2] of the kind 'Bachelors are unmarried men' and 'Bachelors are men'.

However, according to Quine, we have not been successful in formulating a non-circular criterion by means of which we can distinguish the supposed analytic statements from the synthetic ones; nor have we managed to establish a direct connection between each synthetic statement considered in isolation and a complex of verifying or falsifying, confirming or disconfirming, experiences. Therefore, we must conclude that "it is nonsense, and the root of much nonsense, to speak of a linguistic component and a factual component in the truth of any individual statement. Taken collectively, science has its double dependence upon language and experience; but this duality is not significantly traceable into the statements of science taken one by one".[3]

In subsequent literature, the critique of reductionism has been quite widely accepted; however, not all thinkers envisage reductionism as intrinsically connected to the dogma of analyticity, thereby giving up analyticity as well. In defending analyticity, the principle that the best defence is attack has been widely followed, and Quine's position has

---

[1] Quine 1964d: 41.
[2] Quine 1964d: 36f.
[3] Quine 1964d: 42.

been disputed. Since the seventies I have been insisting especially on three aspects of Quine's position which still seem to me particularly problematic.[1] I have emphasised the following difficulties of Quine's conception:
(A) the obviousness of analyticity;
(B) the difficulties of behaviourism and physicalism;
(C) analyticity and reductionism are not "at root identical".

A. The obviousness of analyticity
The first problem concerns the obviousness (acknowledged also by Quine) of the contested dichotomy. Katz referred to this in formulating an operational criterion of analyticity which makes able use of Quine's analysis of the relation between meaning and 'collateral information socially shared'.[2] But, even without considering the efficacy of the test suggested by Katz, there remains the fact – at least for me, difficult to digest – that the denial of any value to the distinction would entail the collapse of the dichotomy between language and theory. As Chomsky once remarked, one cannot doubt "that a person's language and his 'theory' are distinct systems": to do so would mean "that two monolingual speakers of the same language cannot disagree on questions of belief, or that controversy over facts is necessarily as irrational as an argument between a monolingual speaker of English and a monolingual speaker of German".[3] If we want to avoid these conclusions, bordering on the absurd, we must admit the existence of a distinction *in principle* between language and theory, independently of the possibility of individuating it behaviouristically in a non-ambiguous way. This fits in with the common opinion, according to which it is possible to state the same theory in different languages and a plurality of different, even contrasting, theories in the one language.

Therefore, my commitment to the distinction in principle between language and theory depends in the first place upon my unwillingness to go against commonly observed phenomena. These can be described, turning against Quine some of his own felicitous formulations, as follows: "[l]anguage settles the sentences and what they mean; a theory adds, selectively, the assertive quality or the simulation of selective belief. A language has its grammar and semantics; a theory goes farther and asserts some of the sentences".[4]

Commitment to the distinction in principle between analytic and synthetic statements entails some commitment to the traditional philosophical idea that, once the grammar and the semantics of a language have been given, the truth value of some of its sentences – the analytic ones – will depend only upon the characteristics of the language in which they are formulated.

[1] Parrini 1976b.
[2] See Quine 1970: 37-42; Katz 1967, Quine 1967; Katz 1968.
[3] Chomsky 1975: 53f.
[4] Quine 1975a. 309.

As I will explain further on, this is not to say that there exists an absolute boundary, which can be traced on a logical or naturalistic basis, between analytic and synthetic statements, between dictionaries and encyclopaedias. The boundary is certainly fluid. Theory change can provoke language change, and *sentences* which were considered analytic can become synthetic and vice versa. As Sellars said, one of the purposes and effects of science is the change of 'meanings'.[1] What I mean is only that the distinction between language and theory is recognised by common sense. It is presupposed by the verbal behaviour in which we distinguish between disagreement about ways of expressing ourselves and disagreement about beliefs.

Furthermore – as I will explain shortly – the distinction between language and theory can be legitimised at the philosophical level with regard to the epistemic justification of our claims about the empirical adequacy, confirmation and truth value of declarative sentences.

B. The irreducibility of intentionality and the difficulties of behaviourism and physicalism

In the early seventies I showed[2] the intrinsic difficulty of the behaviourist and physicalist position which Quine adopted in order to reject the clarifications of analyticity and the other intensional notions that were being suggested. To be sure, we cannot deny the quasi-circular nature of these clarifications. Intensional notions are a subclass of intentional ones, and it is not possible to define the former (for example, meaning) without presupposing the latter (for example, belief). But if, because of this, we follow Quine in turning our backs on Brentano (and Husserl), rejecting intentionality and intensional concepts,[3] we have to give up the notion of meaningfulness as well. Our assertions would be reduced to "noise-makings" and our thoughts to "*mere* subvocalizations". The only result would be a "mental suicide".

These judgements were expressed by Putnam in an essay of 1981 about the impossibility of "naturalising reason", of "eliminating the normative" from the theory of knowledge. The theory of knowledge cannot be reduced completely to 'naturalised' or 'evolutionary', epistemology,[4] such as the more or less radical theories outlined by Quine, Popper and Lorenz. Putnam is defending the normative, ideal, aspect inherent in the notions of truth, rational acceptability, and so on. But this aspect is closely connected to the use of intensional and intentional concepts.

Furthermore, normativity is trivially connected to the problem of the justification of our cognitive claims, and therefore to the relations between epistemic relativism and the status of meanings, analyticity and

---

[1] Sellars 1967: 288.

[2] See Parrini 1976a, in particular 1976b.

[3] See Quine 1970, especially pp. 220f (on Chisholm's development of Brentano's conception in Chisholm 1957); for Quine's position on Husserl, see Quine 1985: 136f.

[4] Putnam 1983e: 244-7, in particular p. 246

apriority. In discussing this third and last cluster of problems, I will start again from the connection, established in 'Two Dogmas of Empiricism', between analyticity and reductionism. This will also give me the opportunity for an up-to-date theoretical evaluation of the neoempiricist critique of Kantian transcendentalism.

C. Analyticity and reductionism are not "at root identical"
According to Quine, the rejection of reductionism in the name of Duhem's holism entails the denial of the existence of statements completely devoid of empirical content, such as the supposed analytic statements. For those who, like myself, accept analyticity, the inadequacy of reductionism entails, more modestly, theoretical relativism, according to which one can talk about the empirical content of this or that *synthetic statement* only relatively to suitable auxiliary assumptions chosen on the basis of various methodological-axiological considerations.

Consequently, it is possible to use the data of experience as codified in suitable 'protocols', in order to *justify* our evaluations of the empirical adequacy of sentences, of their empirical confirmation or of their truth value, only *relative* to other *contextual assumptions*. These assumptions establish (explicitly or implicitly) the rules of the language that the sentence belongs to (and therefore its meaning), and the general theoretical and / or methodological principles necessary to overcome the systematic ambiguity of observational data. In brief, our practices of epistemic justification depend, not only on the question of observation, which I will consider in the next chapter, but also on the acceptance of a set of analytic, synthetic and methodological principles taken as a priori valid in the given context.[1]

In my view, a point of such philosophical relevance cannot be set aside in the name of a behaviouristic empiricism such as Quine's. Quine's position, as well as presenting internal difficulties, simply excludes all that cannot be reconstructed on the sole basis of the observable verbal behaviour of the speaking and knowing subjects.

4.2. MEANINGS AND EPISTEMIC JUSTIFICATION

By putting intensional notions among the components of epistemic justification, I do not intend to treat 'meanings' as Platonic or mental entities, independent of the cultural processes which language is intimately connected to. Likewise, by granting an epistemic value to the distinctions between analytic and synthetic, a priori and a posteriori, I do not ascribe to them an absolute, ahistorical fixity, independent of our concrete practices of discursive justification.

The questions (1) of linguistic universals and (2) of the existence of meanings as ideal, abstract, entities, are extremely complex, and I will not examine them in detail.

---

[1] This is the thesis I defended in Parrini 1976c, in particular pp. 264-90

(1) I have shown elsewhere that the alleged dependence of the learning of language and of the formation of common sense beliefs upon "biologically necessary" mental structures (Chomsky) does not automatically entail the rejection of the neoempiricist criticism of Kant's theory of the synthetic a priori. The history of science supplies very convincing proofs of our capacity to deeply modify the most well-established linguistic uses, and to develop highly innovative theories under the 'prodding' of experience. Therefore, there is no reason to consider supposed innate forms as 'non-transcendable' limitations of our linguistic and theoretical creativity.[1]

(2) I am confident (this confidence was instilled in me by Giulio Preti) about the possibility of keeping the concessions to Platonism within the boundaries set by the theory of spiritual being elaborated by Hartmann many decades ago.[2] This theory has been appreciated very little, if at all, in many recent discussions on universals. This is a real pity, since it allows us to avoid the Scylla of psychologism without falling into the Charybdis of Platonism. Hartmann's theory recognises a wide autonomy and objective consistency in the world of 'spiritual productions' with respect to the world of the concrete processes of thought; but it does not hypostatise 'spiritual productions' in a Platonic hyperuranium, ontologically totally independent from the real world in which we exist as thinking and judging beings. Although I disagree with Schlick in considering Platonist substantialisations as meaningless, and in considering concepts (and meanings) only as "a kind of shorthand", I believe he was right in denying that concepts and truths have consistency or subsistence "independently of creatures that judge and comprehend".[3]

In my view, meanings and distinctions between analytic and synthetic, a priori and a posteriori, are constituted and modified in a kind of quasi-circular symbiosis with human justificational 'practices', naturalistically existent, and historically and contextually determined. Between a word and its meaning there is no natural connection of a causal or of any other kind, independent of the designative intentions of the user(s) of a certain sound or a certain grapheme. Furthermore, there is no historical fixity in the association between certain phonetic and graphic forms and the meanings they convey, whereby certain sentences must necessarily express certain intensions (*propositions*) rather than others (likewise, there are no meanings or analytic statements such that the linguistic forms that can express them must exist).

Therefore, we can accept Quine's argument that there are no *sentences* whose truth value is not revisable, without having to sacrifice the notions of analyticity and apriority and their epistemic value. We only concede the possibility that sentences can change their meaning and their role. Sentences that expressed analytic propositions can come

[1] See Parrini 1980d, in particular pp. 170-6
[2] See Hartmann 1933.
[3] Schlick 1974: 24f; see 22-7.

to express synthetic propositions (and vice versa). Sentences that expressed contextually a priori synthetic statements can come to express contextually a posteriori synthetic statements (and vice versa). (It is clear that in the first case also the propositional content changes, whereas in the second case it does not).

In my epistemological approach, intensions, the contextual synthetic presuppositions, and the relative distinctions between analytic and synthetic, a priori and a posteriori, do not have an absolute 'foundation' in logic or in nature. Nor do they have an unchangeable ahistorical fixity. Rather, their status is that of historically changeable semantic and epistemic forms; their change and replacement is made possible by the enrichment of conceptual, theoretical and empirical connections, eminently produced by scientific activity, within which they perform their full epistemological role.

Of course, I am aware that terms and concepts often follow complex and blurred rules of use. I am especially thinking of the fact – widely emphasised in contemporary philosophy of science – that most fundamental scientific notions (such as force, mass, energy) are paradigmatic examples of "law-cluster concepts".[1] They are part of a complex 'web' of connections together with many other concepts and principles. This web is generally deeply altered in the course of vast, so-called revolutionary scientific restructurings.

The application criteria of scientific terms and concepts are manifold and often not clearly condensable in a defined set of 'distinguishing marks'. Therefore, we should not expect the distinctions between analytic and synthetic connections, between contextually a priori and contextually a posteriori statements, to be generally "sensed or claimed"[2] by those who are immersed in scientific practice. As if this were not enough, in the next chapter I will argue that understanding the functioning of a descriptive language endowed with extralinguistic reference presupposes that a certain number of its predicates cannot be verbally defined by means of the specification of necessary and sufficient conditions.

Despite this, I am unwilling to conclude that the distinctions between analytic and synthetic, a priori and a posteriori, suitably relativised to a context, retain no epistemological role. These distinctions are legitimised as soon as properly cognitive activity commences. Cognitive activity is something more than purely discriminatory behaviour, governed by the causal mechanisms of stimulus and response. It takes place in what Sellars called "the logical space of reasons, of justifying and being able to justify what one says".[3] The distinctions between analytic and synthetic, a priori and a posteriori are connected to our comprehension of the justificational practices carried out by subjects that make

---

[1] Putnam 1975b, in particular p. 52 (with reference to Wittgenstein's "metaphor of a rope with a great many strands, no one of which runs the length of the rope").

[2] Quine 1970: 57.

[3] Sellars 1968 169.

linguistically formulated cognitive claims, and adduce confirming and disconfirming experiences, described by means of protocol statements at the time believed to be non-problematic.

I am tempted to say that meanings, the analytic and the synthetic a priori, are like linguistic and theoretical coordinates, presupposed by the practices of empirical justification of our cognitive claims, and constituted in relation to these practices. They are symbiotic functions, whose role it is to channel experience so that it may be warrantedly invoked in favour, or against, certain sentences. Therefore, the point I want to make is not the possible psychological and 'expositive' value of these notions. It is their epistemological role, on the level of Kant's *quid juris*. The normativity of knowledge highlights the problem of justification, and warrants the reference to a system of contextually a priori linguistic and theoretical coordinates.

I will describe a simple example: reference to common experience in order to prove the wholesomeness of buttermilk presupposes that 'buttermilk' is intended so as to apply to substances with characteristics that set them apart, let's say, from substances commonly called 'arsenic' (Grünbaum introduces this example in order to criticise "the trivial validity of the D-Thesis").[1] The same can be said about Quine's example concerning neutrinos. One can doubt whether these particles have or do not have mass, and whether the sentence 'Neutrinos lack mass' is analytic or synthetic. But what sense would it have to say that experience proves (directly or indirectly, it does not matter here) that neutrinos – *qua* objects which can be picked out on the basis of criteria including lack of mass – have certain other properties, while denying at the same time the analyticity, or even the truth, of the sentence 'Neutrinos lack mass'?

The progress of experience may lead to dispute the truth of any sentence, even of a sentence which, in a certain context, expressed an analytic statement or an avowedly stipulative linguistic convention, such as the definition of 'momentum' as 'mass times velocity'. By maintaining the epistemological role of the notions of analyticity and apriority, I do not intend to deny such obvious facts. I must only specify that, in a similar circumstance, the meaning of the sentence would be changed: the acceptance of the synthetic a posteriori sentence 'Momentum fails to be proportional to velocity' would depend, as well as upon experience, upon the adoption of a different linguistic and / or theoretical framework.[2]

The same holds in the case of terms which are very ramified and rich in connections, such as 'force' or 'energy'. It would be senseless to set oneself the problem of justifying empirically and theoretically the truth of the sentence 'Force equals mass times acceleration', if one confers definitional status upon Newton's second law. By asking the question, for example, whether experience confirms or disconfirms such a

[1] Grünbaum 1973. 111.

[2] The example is in Quine 1970. 57.

sentence, one would be *eo ipso* excluding that this sentence is used to express an analytic proposition or to state a contextually a priori assumption. On pain of more or less patent absurdity, one would have to refer implicitly or explicitly to a notion of force and to theoretical presuppositions (e.g., Hooke's law), by means of which it were possible to confer a genuinely empirical value to the question whether force equals mass times acceleration.

5. APRIORISM, NEOEMPIRICISM, AND 'NEW PHILOSOPHY OF SCIENCE'

5.1. QUINE AND THE NEOPOSITIVISTS' RELIANCE ON ANALYTICITY FOR REDUCTIONISM

With what I have just said in the conclusion of the previous section, I do not intend to deny the historical relevance of Quine's critique of the two dogmas of empiricism. Even without considering his considerable 'relapses' into philosophy of language, of logic and of mathematics, Quine's critique had the undeniable merit of exposing the logical empiricists' marked tendency to use the distinction between analytic and synthetic in order to ascribe certain very general presuppositions, not easily reducible to hypothetical assertions or to methodological principles, to the linguistic component of science. As I have argued elsewhere,[1] this tendency has very complex historical and theoretical connections with the following: the reductionist formulations of the verifiability principle; Wittgenstein's conception of the tautological character of logic; the neopositivists' insufficient awareness of Duhem's objections to Poincaré's linguistic conventionalism. All this induced the neopositivists to equate the more general principles of science, not endowed with a directly specifiable empirical content (for example, the coordinative 'definition' of congruence at the basis of measurement), with statements devoid of empirical content, that is, with analytic statements interpreted, à la Wittgenstein, as tautologies empty of factual content.

Duhem's holism and Quine's anti-reductionism make it clear instead that an isolated statement, empty of *direct* empirical content (such as, for instance, the principle of inertia or the principle of the rectilinear propagation of light in an homogeneous medium) can have an *indirect* empirical content in conjunction with other statements of the theoretical system of reference.

---

[1] For the justification of the following reconstruction and for the indication of the relevant published and unpublished texts, see Parrini 1980b, 1983; 1987; 1994b, 1993; 1994c, 199+a.

5.2. REICHENBACH VS SCHLICK

Another elucidation is derivable from Quine's critique to the two dogmas of empiricism. If we make use of his objections, not to reject the notions of analyticity and apriority, but to give them a purely relative and contextual value, it becomes possible to gain a better understanding of the differences which, at the beginning of the twenties, divided Schlick and Reichenbach on the relations between the philosophical significance of the theory of relativity and the rejection of the transcendental theory of knowledge.

Schlick, in his critique of Kant, clearly opposed the possibility – upheld by Reichenbach – of separating the constitutive from the apodictic character in Kant's notion of synthetic a priori judgement. At the same time, both Schlick and Reichenbach rejected radical forms of sensationalist empiricism and recognised the necessary presence in knowledge of conceptual assumptions (in the form of coordinative connections) capable of establishing a kind of 'bridge' between the plane of abstract concepts, implicitly defined by the theories' axioms, and that of concrete concepts, which can be characterised empirically by means of ostension. Therefore, the disagreement on Kantianism concerned the epistemological status of such principles of connection.

Reichenbach called them "principles of coordination", and believed that they should be considered constitutive principles of empirical objectivity, analogous to Kant's synthetic a priori principles, except for the character of universality and necessity.

Schlick, instead, followed Poincaré in qualifying them as conventions concerning the linguistic component of science. His declared purpose was the defence of an empiricist position, purportedly intermediate between Mach's sensationalism and Cassirer's logical idealism (Schlick accused Mach, in part wrongly, of neglecting the problem of conceptualisation, and criticised Cassirer for being too faithful to Kant's synthetic a priori). According to Schlick's new empiricism, in science there are only "*hypotheses* or *conventions*",[1] that is, synthetic statements which cannot be considered a priori because they are not apodictic, and conventional assumptions which cannot be considered synthetic because of they are purely definitional or linguistic (analytic).

At a certain point, Reichenbach too 'surrendered' to this conclusion, and he considered his old constitutive principles "coordinative *definitions*", that is – to use Schlick's words – "conventions, in Poincaré's sense".[2] However, initially Reichenbach defended the tie between the notion of coordination and that of constitution. In order to attenuate the contrast with his ally, he tried to reduce the dissent to a 'terminological' question, namely, whether it was appropriate or not to maintain the reference to Kant, when, despite denying the existence of

---

[1] Schlick 1978: 324.
[2] See Reichenbach 1958: 14ff; Schlick 1978: 333.

apodictically certain a priori syntheses, one admitted the necessity of coordinative connections.

What I have said about epistemic relativism shows that the question was not merely verbal. Behind the terminological difference, there was a problem of substance, which came out clearly in the later developments in the philosophy of science. These developments exposed both valid and invalid aspects of the positions of both contenders, according to the 'global' or 'local' point of view from which we look at the epistemic justification of our cognitive claims.

From the 'global' point of view, Schlick was certainly right in deducing from the theory of relativity the philosophical lesson that scientific discourse, considered globally, consists in sentential components expressing only hypothetical synthetic propositions, and sentential components expressing analytic propositions.

However, from the 'local' point of view, Reichenbach had good reasons on his side. By attributing a constitutive value to the "principles of coordination", he proved that he was somehow aware that the epistemic justification of 'local' components of science presupposes reference not only to linguistic-analytic assumptions, but also to theoretical-synthetic presuppositions. This is a striking anticipation of one of the most salient aspects of post-neopositivist philosophy of science.

According to Reichenbach, the "principles of coordination" are distinct from the usual scientific laws and statements in that they make the empirical evaluation of these scientific components possible. Nor does this exempt them from a global empirical comparison with experience. The difference lies in the fact that, while the validity of usual laws and statements depends upon the linguistic-theoretical reference frame and upon specific experiences, the validity of the constitutive principles "does not depend only upon the judgment of particular experiences, but also upon the possibility of the whole system of knowledge": "[i]n contrast to particular laws, they do not say *what* is known in the individual case, but *how* knowledge is obtained; they define the knowable and [...] show the order rules according to which knowledge is obtained". In this sense, they have a constitutive value and they can be considered a priori, provided that it is clear that here "'[a] priori' means '*before* knowledge', but not 'for all time' and not 'independent of experience'".[1]

5.3. REICHENBACH'S 'PRINCIPLES OF COORDINATION' AND KUHN'S 'PARADIGMATIC PROPOSITIONS'

Reichenbach thus anticipated the post-neopositivist idea that in scientific knowledge there operate theoretical principles constituting an integral part of the dominant scientific paradigm, whose task it is to structure the relation between experience and the other synthetic components of the

[1] Reichenbach 1965: 104f

theory. These principles are distinguished from the usual analytic statements in that they do not express purely linguistic or conceptual connections; they differ from ordinary synthetic statements in that they are protected from direct empirical refutation.

In the climate of the 'New Philosophy of Science', Harold I. Brown has called these presuppositions, in homage to Kuhn, "paradigmatic propositions". Such presuppositions "are neither analytic, [...] nor eternal truths", "nor empirical in the usual sense" "exactly because they are protected from straightforward empirical refutation. [...] They constitute an epistemically distinct class in that they do not fit the traditional division of all propositions into a priori and empirical. Rather they are propositions which are accepted as a result of scientific experience but which come to have a constitutive role in the structure of scientific thought".[1]

The individuation of these presuppositions, like the distinction between analytic and synthetic components, far from being trivially evident, is the result of a subtle work of philosophical analysis of theories and scientific arguments. This analysis has proved itself useful also for the developments of science (to be persuaded, it is sufficient to think of all the analytic work which has been done, first on Newtonian physics, and then on relativistic and quantum mechanics – not to mention the current discussions on the foundations of cosmology, biology and psychoanalysis). The idea that this analysis relies on notions and distinctions void of any epistemological significance, which do not capture any relevant aspect of the way in which scientific knowledge is constituted, is clearly contradicted by the connection of these notions and distinctions with the process of epistemic justification of our cognitive claims. Furthermore, in Chapter VI I will show that part of the answer to the problem of the objectivity of knowledge presupposes the use of the distinctions between analytic and synthetic, a priori and a posteriori.

[1] Brown 1979: 105. See also Kuhn's concise elucidations in Kuhn 1983: 566f.

Chapter III

THEORY, OBSERVATION, AND EMPIRICISM

1. RELATIVISM'S THREAT TO EMPIRICAL TESTABILITY

In the previous chapter, I said that in present day philosophy the thrust towards relativism has acquired such momentum as to seriously threaten the empirical and objective character of knowledge. The problem for empiricism is that epistemic relativity endangers the possibility of a genuine, non-circular empirical test of our cognitive claims. If linguistic apparatuses are not simple instruments to report 'what there is out there in the world', and if observation sentences are 'permeated', or 'laden', with theory, we run the risk of not having at our disposal an independent, impartial and neutral empirical touchstone for evaluating our statements.

The question is closely connected with the thesis of the theoretical character of observation, and with the criticisms of the distinction between expressions referring to things considered paradigmatically observable (redness, hardness, spatial and temporal properties, material substances such as water and iron, etc.), and expressions referring to things considered paradigmatically non-observable, or theoretical (mass, kinetic energy, electric fields, atoms, electrons, genes, viruses, wave functions, and so on).

The problem had already been clearly addressed within logical empiricism. This will perhaps be a surprise for those who still believe in the logical empiricists' uncritical faith in an absolutely stable observation language constituting the unalterable basis of the empirical interpretation of theories. I am not only thinking of the now famous controversy over protocols, which saw the contrast between Schlick on the one side and Neurath, Carnap and Hempel on the other; I am also thinking of some less well known aspects of Reichenbach's thought. Reichenbach was a convinced supporter of the idea that there are no unrevisable "conclusive judgements" at any level of scientific discourse. This is clear, as well as in some unpublished texts, in a contribution of his to the controversy between Eddington and Campbell, dating from the early thirties.[1] He certainly did not go as far as to say (as was to be said subsequently) that facts are small theories, and theories are large facts; nevertheless, he maintained firmly that there is no sharp boundary

[1] Reichenbach 1931. The unpublished documents concern the relationship between Reichenbach and C.I. Lewis. See Reichenbach's letter to C.I Lewis of 4.v.1938 [HR-13-49-37], where he expounds the idea that even low level assertions are more or less probable, not true or false; see also Reichenbach's letter to R. Brandt of 10 iii 1951 [HR-37-3-43], in which he says "I agree with him [C. I. Lewis] in a fundamentally empiricist attitude, I differ from him in that I cannot accept any basic statements as absolutely certain, or as terminating judgements, as he puts it".

between facts, laws, and theories. There are only different degrees of probability between experimental facts well established on the basis of solid theories, and experimental facts less well established on the basis of more or less problematic hypotheses. Reichenbach had referred to probability already in an essay of 1929, in order to break the circularity that threatens empirical testability when we affirm the theoretical character of observation.[1]

I persist in the belief that the relations between logical empiricism and the new philosophy of science are much more complex and blurred than they are often still made out to be. However, my present concern is the reconciliation of epistemic relativism, connected with the thesis of the theoretical character of observation, with the possibility of empirically testing our hypotheses and theories.

In recent discussions, the theoretical character of observation has been established following, roughly, two converging lines of argument. I will distinguish them only for convenience of presentation.

The first line has been pursued especially by authors such as Achinstein, Maxwell and Putnam. They criticise the distinction between theoretical and observational language, showing that "the descriptive terms of science generally could be used to refer to both directly observable and nondirectly observable entities and attributes".[2] If we keep to scientific practice, we do not have an important "distinction between two *languages*" (theoretical and observational), but a less binding dichotomy between two "different kinds of uses within the same language".[3]

The second line of argument has been followed principally by the best known champions of the new philosophy of science, starting with Hanson, Kuhn and Feyerabend. Under the influence of ideas of Wittgenstein, Popper, Fleck and Polanyi, they have shown that the meanings of the expressions belonging to the so-called observational vocabulary are no less problematic than those of the expressions belonging to the theoretical vocabulary. Therefore, the expressions of the observational vocabulary cannot constitute an independently accessible key to the understanding of the theoretical terms by means of which theories are formulated.

In order to focus on the possible implications of these lines of argument for empirical testability, I will start again from the semantical conception of observation discussed in the previous chapter. If this conception were right, the objective touchstone of the empirical merits of our hypotheses and theories would be the sentences whose truth value and meaning were logically determinable in a stable and univocal manner on the basis of experience alone. These sentences would express observation statements which would be incorrigible, neutral and invariant with respect to theoretical change. However, criticisms of the kind pre-

[1] See Reichenbach 1978f, in particular sec. 9, 'Physical Fact'.
[2] Suppe 1989: 311.
[3] Hesse 1974: 10.

sented by Feyerabend show that observation does not logically determine the truth value and meaning of pragmatically observational sentences. This automatically raises the problem of the genesis of interpretation, or, more precisely, of what are the relations between constitution of meaning, theoretical elaboration and observation.

Different answers will entail different implications about the empirical testability of hypotheses and theories. Setting aside historical reconstructions, I will outline three abstract theoretical alternatives to the semantic conception. My aim will be to show that the now widely accepted rejection of an absolute logical or naturalistic distinction between theoretical and observational vocabulary does not preclude the possibility of genuine empirical tests of our cognitive claims.

2. THE STRONG THESIS OF THE THEORETICAL CHARACTER OF OBSERVATION

This alternative to the semantical conception of observation is the development to their most radical consequences of the ideas of the new philosophers of science. It affirms the complete dependence of observation on theory. For this reason, it can be called the 'strong thesis' of the theoretical character of observation sentences. Leaving aside the differences between the various authors, this thesis can be condensed into the assertion that the meaning of descriptive terms (even of those used in pragmatically observational statements) is completely determined by the theory they belong to.

Feyerabend, for example, repeatedly declared that the meaning of observational expressions can be determined only by accepted theoretical presuppositions. The latter are to be considered as conventional, and not as synthetic a priori constraints, which, as the positivists rightly pointed out, turned out to be "inconsistent with actual scientific practice".[1] There are no observation sentences associated with a core of objective meaning which is unchangeable and independent of any implicit or explicit subjective assumption. Therefore, the pragmatic theory is right in claiming that observation sentences, devoid as they are of any particular observational meaning, can be separated from theories only pragmatically, not semantically.

Consequently, the traditional formulation of the relation between theory and observation must be reversed. It is incorrect to maintain "that observation sentences are meaningful *per se*, that theories which have been separated from observation are not meaningful, and that such theories obtain their interpretation by being connected with some observation language that possesses a stable interpretation" . The contrary is true: "[t]heories are meaningful independent of observations; observational statements are not meaningful unless they have been connected with theories. [...] It is therefore the *observation sentence* that

[1] Feyerabend 1965: 210.

is in need of interpretation and *not* the theory":[1] "*the interpretation of an observation language is determined by the theories which we use to explain what we observe, and it changes as soon as these theories change*".[2] Following this line, Feyerabend went so far as to deem too half-hearted the thesis that observation concepts are *theory-laden*, asserting that they "are *completely* theoretical".[3]

2.1 THE COLLAPSE OF THE DISTINCTION BETWEEN LANGUAGE AND THEORY AND THE CIRCULARITY OF EMPIRICAL TESTING

It follows from the strong thesis that every theory has its own universe of discourse, which includes the facts and experiences which it posits as its own basis. Theories governed by different principles will not have a common observational ground constituted by observation sentences with the same meaning. Although experience of the phenomenon causally determines the acceptance or the rejection of observation sentences, we are no longer able to see how experience can be used to determine the truth value and the meaning of interpreted observation sentences. If the interpretation of all extralogical descriptive terms is determined by the whole theoretical context to which they belong, then terms belonging to alternative theories will have different meanings, and between such theories there will be no common observation sentences interpreted in the same way. It will be possible to speak neither of invariance of the observational basis with respect to theoretical change, nor of stability, neutrality and incorrigibility of a complex of basic sentences.

This threatens both the role of experience in the choice between alternative theoretical systems, and the existence of an observationally neutral way of deciding between competing theories. Appeal to common intertheoretical empirical data will no longer be possible, since experience of the phenomena will not be able to show the correctness of the interpretation couched in an observation language: "each theory will possess its own experience, and there will be no overlap between these experiences. Clearly, a crucial experiment is now impossible [...] because there is no universally accepted *statement* capable of expressing whatever emerges from observation".[4]

All this raises the well known question of the incommensurability of alternative theories, challenging the idea of a progressive, cumulative, development of scientific knowledge. The feasibility of this kind of progress presupposes adherence to the two conditions of consistency and invariance of meaning in the passage from one theoretical complex to the other. These two conditions are held by Feyerabend (from the essay 'Explanation, reduction and empiricism' (1962) onwards) to be not

---

[1] Feyerabend 1965. 213.
[2] Feyerabend 1981b· 31.
[3] Feyerabend 1983: 50
[4] Feyerabend 1965: 214.

only unrealisable, but even methodologically undesirable, since they might limit the proliferation of theories which constitutes the best guarantee of the growth of science. In order to account for scientific changes, such as the passage from Aristotelian to Galilean, and from Newtonian to Einsteinian physics, we must invoke notions different from sensory experiences and necessary demonstrations; we must refer, for example, to religious, metaphysical and ideological motivations, to rhetorical persuasion, to aesthetic judgements – even to "mystical conversion" or the "gestalt switch", which was emphasised particularly by Hanson and Kuhn.

Although the case for the strong thesis has been cogently put by the new philosophers of science, nevertheless it presents many difficulties, one of which seems to me particularly important. I am thinking of the fact, criticised especially by Shapere,[1] that the champions of the strong thesis have not indicated criteria to establish the precise meaning of the expressions 'to be part of a theory' and 'theory change'. Consequently, they have not explained in a satisfactory manner what is meant by 'to be part of the meaning of a term' and 'change of the meaning of a term' in the passage from one theoretical context to another.

If we consider as integral parts of a theory all the beliefs shared by those operating with a certain language, the theoretical system ends up by incorporating not only *stricto sensu* scientific suppositions, but also common sense persuasions and ideological, metaphysical, religious opinions. This leads to undesirably counterintuitive consequences. The meaning of the theoretical and observational terms used by great scientists, such as Kepler, Newton and Einstein, would depend not only upon their specific hypotheses and theories, but also upon their ideas on the harmony of the universe, on the nature of God, on Spinozism, etc., and it would vary as these vary, despite the obvious disparity of the 'theoretical' spheres concerned.

Nor is the situation any better if we consider the relation between theoretical and intensional changes from the point of view of the more particular, rather than the more general, beliefs. In science, as in everyday life, theoretical changes are going on all the time which "seem too marginal to have an import on the meanings of an expression (for example, the addition of an epicycle, the passage from circular to elliptical orbits, the attribution of a new property to some kind of entity, etc.)".

If we say that these changes represent "changes of meaning", and therefore "changes of experience", it is necessary to compromise ourselves with the "stronger still thesis according to which, for example, a simple extension of the applications of a theory constitutes a difference with respect to the *theoretical context*, and therefore with respect to the meanings of the observation terms" of the theory.

---

[1] Shapere 1984a, in particular 1984b; see also Putnam 1975c: 124f

If, on the other hand, these changes are considered "non pertinent", because they do not constitute "*theory changes*, the problem immediately arises to establish" at which point the "changes *in* the theory" become so "relevant as to constitute changes *of* the theory" capable of having an import on the meaning of the terms.[1]

It is clear that the difficulty I want to underline is connected with the distinction, defended in the previous chapter, between linguistic and theoretical relativism. If we uphold a radical dependence of meaning upon theory, and we do not take care to distinguish between semantic rules and theoretical principles, different meanings come to correspond to different systems of beliefs; again, the distinction collapses between language and theory, between complexes of meanings and complexes of beliefs.

This collapse is connected, in its turn, with the most problematic methodological consequence of the strong thesis: the vicious circularity of empirical testing. If among the beliefs that determine the interpretations of the scientific vocabulary (comprising the observation vocabulary) we include theoretical hypotheses *specific* to a certain domain of research, it follows that the meanings of the terms of that domain (including its observation terms) will come to depend precisely upon theoretical hypotheses specific to that domain: the meanings of the terms of Newtonian physics will depend entirely upon Newtonian theory, the meanings of the terms of relativistic physics upon the theory of special and / or general relativity, the meanings of the terms of Copernican astronomy upon Copernicus' theory and so on.

This idea is strongly supported by general philosophical reasons, such as those connected with the criticism of the semantical conception of observation, as well as by specific arguments in the philosophy of science. Such were those elaborated by Duhem in order to show that in science what counts are not "practical facts", that is, theoretically neutral observations (granting for the sake of argument that such observations do in fact exist), but "theoretical facts", that is, experiences interpreted in the light of a theory. Hanson, Kuhn and Feyerabend have strengthened the conclusion, underlining that experience laden with, or permeated by, scientifically relevant theory is not an observation *interpreted* in the light of the theory, but an observation which is intrinsically theoretical. This is illustrated by the well known gestalt cases of the duck / rabbit, and by Hanson's example of Tycho and Kepler observing the change of the spatial relations between the Sun and the Earth.

2.2. HOLISTIC TESTABILITY AND THE 'IRRATIONALITY' OF THEORETICAL CHANGE

However well it is argued for, this conclusion leaves us with the considerable problem of explaining the possibility of the control that the em-

---

[1] Lanfredini 1988: 29f.

pirical data unquestionably seem to exercise on our theories. If scientifically relevant observation is always an experience permeated with theory, and if the presupposed theory is the one we want to evaluate empirically, testing seems to be viciously circular, and it becomes difficult to understand how our theories can clash with adverse data. The strong thesis, thus interpreted, conflicts with the fact that we hold verifiable empirical expectations which are often frustrated (the well known stubbornness of data!).

Arguing against Le Roy's "nominalism", Poincaré pointed out rightly that, although it is possible to treat every law and scientific statement as a definition, unassailable by experience, it will not be possible to reduce the whole of science to a pure conventional construction. It is true that every empirical law can be broken up into a conventional principle and a law; but "however far this partition be pushed, there will always remain laws"[1] with a recognisable and testable empirical content.

The objectivity of empirical testing seems to rest primarily on observation's capacity to clash with a pre-existing horizon of expectations, that is, in Reichenbach's terms, on the (at least relative) independence of perception from reason and its theoretical constructions.[2]

This role of empirical data is recognised also by the new philosophers of science. By contrast with staunch empiricists, the new philosophers of science have certainly aimed at attenuating the import of experience in the evaluation of theories; but they were neither able, nor willing, to deny it completely. They have opted for a position on the whole in line with the outcome of Duhem's criticism of crucial experiments, which endorses the empirical falsifiability and confirmability of theoretical systems globally considered.

Kuhn's position constitutes a particularly instructive example. He insists on the idea that cognitive activity presupposes the interpretation of nature by means of a set of "conceptual boxes". On the other hand, he also maintains that "nature cannot be forced into an arbitrary set of conceptual boxes. On the contrary, the history of proto-science shows that normal science is possible only with very special boxes, and the history of developed science shows that nature will not indefinitely be confined in any set which scientists have constructed so far".[3] If this were not the case, Kuhn would not have been able to talk about the empirical "anomalies" which affect paradigms (understood as "disciplinary matrixes").

The Duhemian spirit of this aspect of the question is well expressed in a neglected remark by Lakatos: "[i]t is not that we propose a theory and Nature may shout *no*; rather, we propose a maze of theories, and Nature may shout *inconsistent*".[4] Nature can, that is, induce us to uphold "a 'factual' statement couched in the light of one of the theo-

---

1 Poincaré 1946c: 335.

2 See Reichenbach 1978e: 38.

3 Kuhn 1987b: 263.

4 Lakatos 1987: 130, capital letters substituted by italics.

ries involved, which we claim Nature had uttered and which, if added to our proposed theories, yields an *inconsistent system*".[1] Feyerabend himself explicitly recognises the possibility of *refuting* "[i]ncommensurable theories [...] by reference to their own respective kinds of experience" – though he specifies parenthetically that "in the absence of commensurable alternatives these refutations are quite weak".[2]

In the light of these considerations, we can better appreciate the point of an important criticism, which especially Achinstein and Suppe have been reiterating since the end of the sixties, against the new philosophy of science, and in particular against Feyerabend. If we maintain that "[t]he meaning of *every* term we use depends upon the theoretical context in which it occurs",[3] then "[e]very statement of a theory is constitutive of the meanings of the terms of the theory and of the observation language"; and "this makes science analytic".[4] Feyerabend's conception seems thus to imply that there are not even internal and limited empirical refutations. It "seems to lead to the position that theories are self-certifying and thus any global theory is as acceptable as any other".[5]

As a matter of fact, the objection, thus formulated, neglects the implications of the distinction between characteristic and interpretation. As Feyerabend explicitly said, to maintain that "there is no universally accepted *statement* capable of expressing whatever emerges from observation", does not entail the denial of the existence of "*human experience as an actually existing process*" which "causes the observer to carry out certain actions, for example to utter sentences of a certain kind. Not every interpretation of the sentences uttered will be such that the theory furnishing the interpretation predicts it in the form in which it has emerged from the observational situation. Such a combined use of theory and action leads to a selection even in those cases where a common observation language does not exist."[6]

This shows that the criticisms of those who (like Achinstein and Suppe) accused Feyerabend's conception of being patently incompatible with the potential conflict between experience and theory, were wrong, or at least wrongly formulated. The distinction between characteristic and interpretation allows us to maintain that the meanings of terms (including observation terms) are radically permeated by theory, without disregarding "the possibility that an observer may be caused by experience to reject an observation sentence which is a logical consequence of the theory".

---

1 Lakatos 1987: 130n.
2 Feyerabend 1987: 227.
3 Feyerabend 1965: 180, italics added.
4 Suppe 1989: 315, with reference to Achinstein 1968: 94-7.
5 Suppe 1989: 316.
6 Feyerabend 1965: 214; see also p. 179f.

But, since "the interpretation of the [observation] *sentence* provided by that theory makes the rejected *statement* analytically true",[1] it is possible to reformulate Achinstein's objection, freeing it from its original defect: "Feyerabend's view does allow the conflict between theory and observation, hence for the observational test of theories; but such conflicts must take the form of rejecting analytically true assertions on the basis of experience under circumstances where one knows that the rejected assertion is a logical consequence of the theory. One who did this while accepting Feyerabend's meaning doctrines (and their consequences) would have to be rejecting *on the basis of experience* a proposition which one *knew* to be analytically true. Experiential falsification thus is possible on Feyerabend's view but involves *fundamentally irrational behavior*".[2] The "anything goes" view appears to be largely the methodological counterpart of the intrinsic irrationality that Feyerabend attributes to theoretical change.

In Feyerabend's perspective, it is practically impossible to understand how experience can play an intertheoretical role, refuting a complex of non-contradictory statements, and confirming an alternative theoretical system. As Shapere said, "Feyerabend's kind of experience is altogether *too weak*, in its pristine, uninterpreted form, to serve as grounds for 'removal' of any theory; and his view of meaning is *too strong* to preclude the possibility of *any* interpretation whatever of what is given in experience".[3]

3. THE WEAK THESIS OF THE THEORETICAL CHARACTER OF OBSERVATION

3.1. THE PARADOX OF CATEGORISATION

The manifest defects of the strong thesis do not reinstate the plausibility of the semantical conception of observation. Criticisms of this conception are so many and so weighty, that it is no longer possible to ascribe to observations alone, with no reference to theoretical presuppositions, the capacity of logically determining the meaning and truth value of part of our linguistic expressions – all the more if we consider that in science we need, rather than generic experience, scientifically relevant experience.

We end up by running into an apparently paradoxical problem. Drawing inspiration from Lewis's position on the observational basis (a position already criticised by Reichenbach because of a certain 'absolutism'), Scheffler formulated the question in general philosophical terms, stating the *paradox of categorization*: "[i]f my categories of

---

[1] Suppe 1989: 316, italics added.
[2] Suppe 1989: 316.
[3] Shapere 1984b: 77; for Suppe's remarks on Shapere's position, see Suppe 1989: 316-9.

thought determine what I observe, then what I observe provides no independent control over my thought. On the other hand, if my categories of thought do not determine what I observe, then what I observe must be uncategorized, that is to say, formless and nondescript – hence again incapable of providing any test of my thought. So in neither case is it possible for observation, be it what it may, to provide any independent control over thought. [...] Observation contaminated by thought yields circular tests; observation uncontaminated by thought yields no tests at all".[1]

It is not easy to overcome the paradox, since the defects of the two extreme positions (the semantical conception and the strong thesis) are not sufficient to prove the possibility of reconciling the theoretical contamination of observation sentences with the non-circular empirical testability of hypotheses and theories.

3.2. AN ALLEGED SOLUTION: THE DISTINCTION BETWEEN CATEGORIES AND HYPOTHESES

In their search for a solution, a number of philosophers, especially Kordig[2] and Scheffler, inclined towards a second alternative. This alternative was to some extent already present in Popper's idea of the different levels of universality of scientific laws, and of the different degrees of dispositionality and theory-ladenness of expressions such as 'red' and 'soluble' on the one hand, and 'quark' and 'neutrino' on the other. This second alternative can be called the 'weak thesis' of the theoretical character of observation, since, while asserting the theoretical contamination of the observational vocabulary, it is formulated so as to avoid the danger that such contamination may depend not simply upon a generic theory, but upon the specific theoretical hypotheses that we want to test.

In Scheffler's version, the weak thesis overcomes the circularity objection by introducing a kind of stratification into the undifferentiated concept of theorisation: we make a distinction between *categories* and *hypotheses*. The concept of theorisation is connected to:

(1) a more or less structured complex of extremely general categories, whose purpose it is to order experience by spelling out the modes of identification and classification of the elements that we want to refer to. "Lacking such order altogether, we may, indeed, aptly be described as facing an undifferentiated chaos, since we lack the very recognition of things – that is to say, we do not individuate and separate items as objects of reference";

(2) specific expectations, beliefs, hypotheses, about how in fact the elements thus identifiable and classifiable will adapt to the "pigeonholes" presupposed by the categorial system.

---

[1] Scheffler 1976: 13f.
[2] Kordig 1971.

The categorial system, therefore, turns out to be "compatible with alternative, and indeed conflicting hypotheses: that is, having adopted a given category system, our hypotheses as to the actual distribution of items within the several categories are not prejudged. Conversely, the same set of hypotheses may be formed compatibly with different category systems: the categories specifically referred to in these hypotheses may belong, as a common possession, to the different systems in question. Simply to set up an alphabetical filing system for correspondence is not yet to determine how tomorrow's correspondence will need to be filed. Conversely, to guess that the next letter will need to be filed under 'E' or 'L' is a prediction that may be made whether or not we have a place for 'X' in our system. [...] Without a vocabulary and grammar, we can describe nothing; having a vocabulary and grammar, our descriptions are not thereby determined. [...] Categorization provides the pigeonholes; hypothesis makes assignments to them".

This distinction between categories and hypotheses would seem to allow a plausible solution of the paradox of categorisation and of the circularity of empirical testing. "To say that a categorization is independent of – in the sense that it does not prejudge – the particular hypotheses expressible through it means that observation may be conceived as thoroughly molded by such categorisation and yet equally independent of these hypotheses. We simply have a false dichotomy in the notion that observation must be either a pure confrontation with an undifferentiated given, or else so conceptually contaminated that it must render circular any observational test of a hypothesis". Therefore, it is possible to have "alternative assignments stipulated by different hypotheses [...] with the same categorization"; and it is possible to decide between conflicting hypotheses without having "to decategorise and unstructure our thought".[1]

This can be expressed in the terms used in discussing the strong thesis. Only interpretative or categorial theories play a role in fixing the meaning of the observational vocabulary. Theoretical hypotheses, by contrast, do not contribute to determining those meanings, and can vary relatively to the categorial theory without their variation influencing the meaning of the observation terms. Therefore, theoretical hypotheses can be subjected to non-viciously circular empirical tests. There are no purely observational terms, devoid of theoretical contamination. But this does not rule out the possibility of empirical tests, which are to some extent objective, even if relative to the background interpretative theory.

I think that the distinction between categories and hypotheses is warranted. In recent literature works have appeared which try to elaborate a logically rigorous notion of categorial scheme and of the associated functions of individuation and classification (it is sufficient to think of the important contributions of Körner, who speaks also about rela-

---

[1] Scheffler 1976: 38f

tively a priori synthetic statements).[1] The distinction is analogous to that between language and theory traced in the previous chapter. In Scheffler's perspective, the complex of categories and the complex of hypotheses are two distinct systems, in the sense in which, according to Chomsky's criticism of Quine, the linguistic system of meanings and the system of beliefs are in principle distinct: as a language allows the formulation of a manifold of different, even incompatible, beliefs, likewise a categorial system allows the formulation of a manifold of different, even incompatible, hypotheses.

Unfortunately, I do not believe that the distinction in principle between categories and hypotheses is sufficient to overcome the threat of the circularity of empirical testing. As in the case of the dichotomies between the complex of meanings and the complex of beliefs, and between analytic and synthetic sentences, this distinction too cannot be made on a logical or naturalistic basis. None of these distinctions has an absolute value, and can be considered invariant with respect to scientific change. But, if we confer a merely contextual value on Scheffler's dichotomy too, we must consider the possibility that the change of hypotheses will bring with it a change of categories; that, in the passage from one theoretical context to another, also the categorial scheme may change. And this makes it impossible to use the distinction as a means to neutralise the threat of the circularity of empirical testing.

Scheffler's solution hinges on the presupposition that the intensions (or at least the extensions)[2] of observational expressions and of the connected categorial expressions do not change along with change of theories. That is, the intensions (or extensions) of observational expressions are *necessarily invariant* with respect to alternative theories, or, as required by the semantical conception of observation, the intensions (extensions) of observational expressions are *neutral* and *stable* with respect to the change of hypotheses.

However, since the fixity of the categorial system is not granted, objections can be raised against this suggestion that are symmetrical to those raised against the traditional conception. It is possible to dispute the scientific relevance of a scientifically neutral, even if categorised, datum. It is possible to question the methodological desirability of observational objectivity and neutrality. Most of all, it is possible to deny that the meanings of categorial expressions and of the connected observational expressions remain certainly unvaried with the change of the theoretical hypotheses.

For example: would it be legitimate to affirm that the sentences concerning spatial and temporal specifications have maintained their meaning unaltered in the course of the passage from classical to relativistic physics? In 1920, Campbell still considered judgements about the temporal order of events as the most stable and incontrovertible judgements one could imagine; but in 1905 Einstein had shown that the

---

[1] Körner 1974.

[2] See also Smith 1981: 7f.

determination of simultaneity at a distance, and the connected judgements, depend upon theoretical assumptions concerning the physically admissible limits of velocity.[1] As Reichenbach said, back in the twenties, discussing the philosophical implications of the theory of relativity, "physics creates new categories which cannot be found in traditional dictionaries".[2]

And, if there are no such things as categorial systems fixed once and for all, which can be individuated with precise logical or naturalistic criteria, there is no reason to suppose that the change of hypotheses leaves unaltered the meaning of the observation terms relevant to the evaluation of the empirical adequacy of those very hypotheses.

4. THE NETWORK MODEL

Despite these difficulties, I do not think we should resign ourselves to the vicious circularity of empirical inference and to a drastic reduction of the role of experience in scientific knowledge. In the last twenty years, a number of empiricist answers have been suggested to the questions raised by the new philosophy of science. It is sufficient to think of Bas van Fraassen's "constructive empiricism", of Dudley Shapere's "contingent empiricism", of Allan Franklin's and Ian Hacking's "new experimentalism", of the new versions of sophisticated inductivist methodologies. (Also Harold I. Brown has set forth ideas that "[t]o some degree [... sound] like a return to classical empiricism".)[3]

Among these various approaches, my preference is for the network model propounded by Mary Hesse in a number of works since the beginning of the seventies. One of her most important goals is that of restoring the possibility of empirical testability without denying the full theory-dependence of observation. The main reason for my choice is the fact that this "new empiricism" – fully developed in *The Structure of Scientific Inference* (1974) – makes a very able use of Duhem's and Quine's ideas which are incorporated in my relativist and contextual conception of the distinctions between analytic and synthetic, a priori and a posteriori.[4] In the following pages I will illustrate those aspects of the network theory which are strictly relevant to overcoming the problem of the circularity of empirical testing.

The network model constitutes a third alternative to the semantical conception. It differs from the semantical conception, and from the weak thesis, because it does not claim to isolate a vocabulary whose ex-

---

[1] The reference to Campbell is in Hesse 1974: 20.

[2] Reichenbach 1978e: 34.

[3] Brown 1987: 219.

[4] See Hesse 1974, ch. I, 'Theory and Observation' (pp. 9-44); see also Hesse 1980, in particular ch. II, sec. 3, 'Theory and Observation' (pp. 63-110). For a detailed account and defence of the network model in relation to the problems raised by the New Philosophy of Science, see Lanfredini 1988.

pressions have an interpretation which is univocally determined by empirical observation or by their relation with a background categorial theory. Internally, there is no distinction between theoretical and observational language based upon specific logical or natural properties belonging to certain expressions and not to others. The network model differs from the strong thesis in that, although it does not require the necessary invariance of observation sentences with regard to theoretical change, it does not exclude a priori a relatively neutral empirical testability of our hypotheses and theories, even of competing ones.

The network conception removes all absolute value, naturalistically or logically specifiable, from the distinction between theoretical and observational expressions, conferring upon it an exclusively pragmatic and relative status, analogous to that which, in my view, should be accorded to the dichotomies between analytic and synthetic, a priori and a posteriori.

The contextual value of these dichotomies was argued for by means of an analysis of the process of epistemic justification of our cognitive claims. This analysis highlighted the dependence of the justification of our cognitive claims upon relatively a priori analytic and synthetic presuppositions. The contextual value of the dichotomy between theoretical and observational vocabulary will depend instead upon the analysis of the epistemological presuppositions at the basis of the functioning of descriptive expressions endowed with an extralinguistic reference. This functioning concerns both the ways in which such expressions are learnt and introduced by the speakers, and the ways in which they function in relation with experience and with other components of the same symbolic apparatus.

A. Linguistic learning and the logical primacy of repetitions

It does not seem possible to understand the process of the formation and learning of a symbolism endowed with an extralinguistic reference, if we do not start by acknowledging that "some predicates are initially learned in empirical situations in which an association is established between some aspects of the situation and a certain word".

Leaving aside the psychological and linguistic dimensions of association, the epistemologically relevant point is the following: "[s]ince every physical situation is indefinitely complex" and "[s]ince every situation is in detail different from every other",

(a) the capacity to identify "the particular aspect to be associated with the word" choosing it out of "a multiplicity of other aspects", and

(b) the capacity to use the word correctly again also in situations which are new with respect to those in which it was originally learnt,

both presuppose the possibility of recognising "degrees of physical similarity and difference [...] between different situations".

In the light of some observations by Wittgenstein, we must certainly concede that "physical similarity [...] is not *sufficient*" to explain the functioning of words with extralinguistic reference, and "that not all referring words need to be introduced in this way". However, we must also recognise that if *no* such words were introduced in the way just specified, it would become impossible "to conceive how an intersubjective descriptive language could ever get under way". Therefore, we are justified in assuming that "the applicability of a language containing universal terms" presupposes that at least "*some* of these terms" are tied to "primitive causal recognitions of physical similarities".[1]

In the history of epistemology and philosophy of science, another important objection has been raised against the thesis of the "logical primacy of repetitions". Popper, for example, repeatedly accused it of emptiness: since evaluations of similarity presuppose the adoption of a point of view, "'with a little ingenuity' we could always find similarities in *some* same respects between all members of any finite set of situations. That is to say, 'anything can be said to be a repetition of anything else, if only we adopt the appropriate point of view'". As a matter of fact, this kind of objection constitutes one of the classical arguments against the empiricist conception of the formation of concepts (think, for example, of Husserl and Cassirer). Nor can we deny that it has a certain validity.

But it is certain that, if we stopped at this consideration, we "would make the learning process in empirical situations impossible": "however finitely large the number of presentations of a given situation-aspect, that aspect could never be identified as the desired one out of the indefinite number of other respects in which the presented situations are all similar. It would, of course, be possible to eliminate some other similarities by presenting further situations similar in the desired respect but not in others, but it would then be possible to find other respects in which all the situations, new and old, are similar – and so on without end".[2]

The behaviourist theory of "a sort of prelinguistic quality space" can help us out of this difficulty. This theory talks about "a prior tendency to weigh qualitative differences unequally", constituting the presupposition of learning mechanisms of the stimulus-response kind.[3] In this perspective, it is possible to reinterpret Popperian objections as a proof of the fact "that the physics and physiology of situations already give us some 'point of view' with respect to which some pairs of situations are similar in more obvious respects than others, and one situation is more similar in some respect to another than it is in the same respect to a third".

According to Hesse, Popper has somehow obscured the question "by speaking as though, before any repetition can be recognized, we have to take thought and *explicitly* adopt a point of view. If this were so, a regressive problem would arise about how we ever learn to apply the predicates in which we explicitly express that point of view. An immediate consequence of this is that there must be a stock of predicates in

---

[1] Hesse 1974: 11f; for Hesse's position on Wittgenstein, see also p. 46.
[2] Hesse 1974: 13; the reference is to Popper 1977b, appendix *x, p. 422
[3] Quine 1970: 83.

any descriptive language for which it is impossible to *specify* necessary and sufficient conditions of correct application. For if any such specification could be given for a particular predicate, it would introduce further predicates requiring to be learned in empirical situations for which there was no specification".

Popperian objections therefore should not induce us to reject as empty learning based on repetition, but to admit "that the primary process of recognition of similarities and differences is necessarily *unverbalizable*" – of course only the *primary* one is unverbalisable, "because it may be perfectly possible to give empirical descriptions of the conditions, both psychological and physical, under which similarities are recognized". The point is only that "such descriptions will themselves depend on further undescribable primary recognitions".[1]

B. The primacy of repetitions does not yield pure observation predicates

These preliminary considerations on linguistic learning and similarity are of great importance in order to overcome the difficulties connected with the traditional way of conceiving the distinction between theoretical and observational language. The fundamental point we must not lose sight of is "that the primary process of classifying objects according to recognizable similarities and differences" attributes a role to empirical situations in the learning of some descriptive expressions, but it does not because of this provide the independent observation predicates required by the semantical conception of observation.

Prima facie, one could think the contrary. But it becomes clear that this impression is a mistake, as soon as we recall the *non-transitivity* of relations of similarity and difference. If two objects a and b (for example, two shades of colour) can be inserted in the class indicated by the predicate P, because they are "similar to some degree" with respect to that predicate, and if the object c (a third shade of colour) is similar to b in the same degree, this does not entail that c is similar to a in the same degree as it is to b, or even that it is similar to a.

Two important consequences follow from this:

(1) some objects are "more 'central' to the P-class than others" (that is, in the class of objects which are judged similar with respect to the predicate P);

(2) "the process of classifying objects by recognition of similarities and differences is necessarily accompanied by some loss of (unverbalizable) information. For if P is a predicate whose conditions of applicability are dependent on the process just described, it is impossible to *specify* the degree to which an object satisfies P without introducing more predicates about which the same story would have been told. Somewhere this potential regress must be stopped by some predicates whose application involves loss of information which is present to recognition but not verbalizable".

---

[1] Hesse 1974: 13.

It is therefore possible to show that "the primary recognition process, though necessary, is not sufficient for classification of objects as *P*, and [that] the loss of information involved in classifying leaves room for changes in classification to take place under some circumstances. Hence primary recognitions do not provide a stable and independent list of primitive observation predicates".[1]

C. The primacy of repetitions and the genesis of theoretical predicates

In order to understand this, we must realise the mistakenness of another idea that might arise spontaneously. We are generally prone to thinking that the method of learning by means of direct empirical associations applies only to predicates such as 'red', 'round', 'ball', etc. Certainly, it applies primarily to predicates of that kind; but in the network model nothing excludes such a possibility even for the introduction of words such as 'chair', 'dinner', 'mama', and even such as 'rule', 'game', 'stomach ache', 'heartache', 'atom'. As a matter of fact, however, if we use only the method of direct association with empirical situations, we do not manage to master completely the descriptive expressions, of whatever kind they may be.

The point is the possibility of treating symmetrically all the descriptive words of the natural language, both those traditionally considered theoretical and those traditionally considered observational: each word can be introduced and learnt by means of the method of direct empirical association; and each word can be inserted, as the language develops, in general statements commonly accepted to be true, such as: 'Balls are round', 'In summer leaves are green', 'Eating unripe apples leads to stomach ache', and so on. In this context, "[i]t matters little whether some of these" *laws* will subsequently be called 'analytic statements', while others ("perhaps most") will be considered 'synthetic'.

"It is not necessary, either, that every such law should be *in fact* true, only that it is for the time being accepted as true by the language community." As a matter of fact, with the development of knowledge "any one of these laws" could be rejected as "*false* (although not all could be [rejected] at once)".[2]

For the present, it is important to note that "[m]aking explicit these general laws is only a continuation and extension of the process already described as identifying and reidentifying proper occasions for the use of a predicate by means of physical similarity. For knowledge of the laws will now enable the language user to apply descriptions correctly in situations other than those in which he learned them, and even in situations where nobody could have learned them in the absence of the laws – for example, 'stomach ache' of an absent individual known to have consumed a basketful of unripe apples, or even 'composed of diatomic molecules' of the oxygen in the atmosphere. In other words,

[1] Hesse 1974: 13f.

[2] Hesse 1974: 14f, with slight modifications.

the laws enable generally correct inferences and predictions to be made about distant ('unobservable') states of affairs".[1]

Furthermore, with the enrichment of expressive tools, it is possible to coin and introduce new terms by means of analogies, metaphors, and theories elaborated by making use of known scientific laws, and by means of the invention of myths. In order to describe the entities, one adds to some predicates already present in the language other predicates introduced with reference to mythical situations and mythical laws. "In other words we build up a network of predicates and laws partly imaginatively, but not in such a way as to contradict known laws, as in a good piece of science fiction". The procedure is analogous to that by means of which, according to Black, "a poet builds up a web of imagined associations within the poem itself in order to make new metaphors intelligible".[2] In this way, it becomes possible to mint and introduce entirely new words into circulation – words about which it cannot be said that they are in principle inapplicable directly to experience, since, with the progress of knowledge, the conditions for their direct empirical application might arise.

D. The holistic character of change

With the progressive enrichment of the complex of the relations between expressions and expressions, and between expressions and empirical situations, the 'fabric' of predicates and of their more or less strictly nomological connections becomes complex and structured enough "to allow for the possibility of internal misfits and even contradictions" which make changes necessary. This possibility can occur in many ways. These are all philosophically instructive, because they show the significant dependence of the changes upon considerations regarding the system of laws and empirical associations in its entirety.

(1) In some cases, "[i]t may happen that some of the applications of a word in empirical situations turn out not to satisfy the laws which are true of other applications of the word. In such a case, since degrees of physical similarity are not transitive, a reclassification may take place in which a particular law is preserved in a subclass more closely related by similarity, at the expense of the full range of situations of application which are relatively less similar. An example of this would be the application of the word 'element' to water, which becomes incorrect in order to preserve the truth of a system of laws regarding 'element', namely that elements cannot be chemically dissociated into parts which are themselves elements, that elements always enter into compounds, that every substance is constituted by one or more elements, and so on".

(2) In other cases, "the range of applications [of a word] may be widened in conformity with a law, so that a previously incorrect application becomes correct. For example, 'mammal' is correctly applied

[1] Hesse 1974: 15.

[2] Hesse 1974: 32; and Black 1962· 43, cited in Hesse 1974: 32.

to whales, whereas it was previously thought that 'Mammals live only on land' was a well-entrenched law providing criteria for the correct use of 'mammal'".[1]

Both examples display the holistic character of the considerations guiding change. In the second case, "it is not adequate to counter with the suggestion that the correct use of 'mammal' is *defined* in terms of animals which suckle their young, for it is conceivable that if other empirical facts had been different, the classification in terms of habitat would have been more useful and comprehensive than that in terms of milk production".

As for the first case, "it cannot be maintained that it is the *defining* characteristics of 'element' that are preserved at the expense of its application to water because, of the conditions mentioned, it is not clear that any particular one is, or ever has been, taken as *the* defining characteristic; and since the various characteristics are logically independent, it is empirically possible that some might be satisfied and not others. *Which* is preserved will always depend on what system of laws is most convenient, most coherent and most comprehensive".

Moreover, the idea that the definition is what decides the correct application clashes with the non-eliminable presence of "a large number of predicates for which *no* definition in terms of necessary and sufficient conditions of application can be given". For example, in a case such as that of 'whale' nothing prevents the desire to "preserv[e] a particular set of laws" from inducing us to set aside a primary recognition of similarity between whales and the members of the set of fish.[2]

4.1 THE PRAGMATIC AND RELATIVE DISTINCTION BETWEEN THEORY AND OBSERVATION

So, the network model rests on the recognition of two salient aspects of descriptive expressions with extralinguistic reference: (1) none of them can work exclusively by means of empirical associations; (2) even those which are introduced and learnt by means of linguistic associations are in principle susceptible to empirical applications.

(1) The validity of the first point is clearly illustrated by a predicate such as 'red', traditionally considered a paradigmatic case of observational expression. In many areas of contemporary philosophy, ranging from Popper's discussion of Fries' trilemma, to Wittgenstein's arguments against a private language, it has been shown that the intersubjective use of a linguistic symbolism requires the functioning of the predicates to depend upon accepted laws. "[I]n using a public language, the correctness of any application of a predicate in a given situation must in principle be capable of intersubjective test. [...] Now if my careful response of 'red' to each of a set of situations were all that were

---

[1] Hesse 1974: 15.
[2] Hesse 1974. 15f.

involved in my correct use of 'red', this response would not be sufficient to ensure intersubjectivity. It is possible, in spite of my care, that I have responded mistakenly, in which case the laws relating 'red' to other predicates can be appealed to in order to correct me (I can even correct myself by this method): 'It can't have been red, because it was a sodium flame, and sodium flames are not red'. If my response 'red' is intended to be an ingredient of a public observation language, it carries at least the implication that disagreements can be publicly resolved, and this presupposes laws conditioning the function of 'red'. If this implication is absent, responses are mere verbal reflexes having no intersubjective significance (unless of course they are part of a physiological-psychological experiment, but then I am subject, not observer)".[1]

(2) Also the second point holds true. There is no doubt that, in science and in language in general, there are descriptive expressions with extralinguistic reference, which are 'anchored' mostly to verbal connections (analogical and metaphorical constructions, theoretical elaborations, poetic and mythological inventions, etc.). And it is equally certain that some such expressions are initially referred to non-directly observable 'theoretical entities'. However, we cannot set limits a priori to the progress of knowledge and to our capacities of producing new experimental conditions. Accordingly, it will not be possible to rule out that even such words may be employed in directly observable empirical situations: "once a theory is accepted and further experimental evidence obtained for it, predicates may well be ascribed directly to previously unobservable entities, as when genes are identified with DNA molecules visible in micrographs or when the ratio of mass to charge of an elementary particle is 'read off' the geometry of its tracks in a magnetic field".[2]

All this makes a distinction between *absolutely* observational and *absolutely* theoretical terms completely illegitimate. Therefore, within the network model there is no place for a "set of observation predicates whose interrelating laws are absolutely invariant to changes in the rest of the network of laws", and whose meanings are completely stable with respect to theoretical and conceptual change. Two particularly perspicuous examples are constituted by the term 'red' and by the term 'simultaneous': in astrophysics, 'red' has taken on a relational value which it does not have in common use; the criteria of application of the term 'simultaneous' have undergone a deep transformation in the passage from classical to relativistic physics.

So, for the network model "no feature in the total landscape of functioning of a descriptive predicate is exempt from modification under pressure from its surroundings". After Neurath and Quine, it has become trite to repeat that "any empirical law may be abandoned in the face of counterexamples". But the assertion is no longer trite "when the

[1] Hesse 1974: 17f, with reference to Popper 1977a: 44f, and to Wittgenstein 1992: sec. 258 ff.

[2] Hesse 1974: 29.

functioning of every predicate is found to depend essentially on some laws or other and when it is also the case that any 'correct' situation of application – *even that in terms of which the term was originally introduced* – may become incorrect in order to preserve a system of laws and other applications".[1]

By contrast with what happened in the standard conception, in the network model there is not a 'vertical' dualism between the theoretical and the observational levels, entailing the necessity of introducing adequate correspondence rules. The network model generates an 'horizontal' view of scientific knowledge, analogous to that found in Quine's metaphor of the field of forces.[2] Scientific theories are seen as a complex of more or less well entrenched laws and lawlike implications, ranging from the most consolidated principles of physics to the most uncertain 'suggestions' endowed with an extremely weak empirical backing. The distinction between theoretical and observational vocabulary has a purely pragmatic, relative and contextual value. Observational expressions progressively qualify themselves as such, distinguishing themselves from theoretical expressions only in virtue of their being "better entrenched" than others. Such is the case, for example, with 'red' with respect to 'ultra-violet', of 'lead' with respect to 'π-meson'. But this is interpreted not as a logical truth "about the *a priori* features of a peculiar set of predicates", but as "a *factual* truth about the relative invariance of some empirical laws to increasing empirical information". The better entrenched words are only less prone than others "to change of function in ordinary discourse"; therefore, they reveal less than others do "the speaker's commitments to a system of laws" or "his relative ignorance" of general connections.

Thus conceived, the distinction between theoretical and observational vocabulary "does not entail that any predicate is *absolutely* entrenched, nor that any subsystems of predicates and the laws relating them are immune to modification under pressure from the rest of the system". Certainly, the distinction aims towards individuating some predicates (the 'observational' ones) that, compared with others (the 'theoretical' ones), "refer to aspects of situations more *directly observable*"; but this only in the sense that the function of these predicates is 'anchored' to empirical situations in a more obvious way than it is to laws.

In this way, the distinction between theoretical and observational vocabulary can explain many epistemically interesting features of our language, starting with the existence of "predicates that are learnable and applicable in a *pragmatically* simpler and quicker manner than others", and which constitute the instruments by means of which other expressions "are *anchored to the empirical facts*". It can do this, however,

---

[1] Hesse 1974: 16f.

[2] See also Quine 1990. 8.

without the pretension of translating such characteristics into absolute properties independent of theoretical and cognitive change.[1]

The main philosophical virtue of this model is that it allows us to follow the new philosophers of science in speaking of the 'dependence on theory', or theory-ladenness, of all the descriptive predicates, without falling into *radical conventionalism*, which, neglecting the empirical content inherent in the primitive recognitions of similarity, maintains "that any law can be assured of truth by sufficiently meddling with the meanings of its predicates".[2] Since without such primitive recognitions of similarity there would be no descriptive language endowed with extralinguistic reference, it will not be possible to infringe all the recognitions in order to preserve a given law. Likewise, it will not be possible to impose any theory upon any set of facts, ignoring the conditions of coherence, simplicity and convenience of the system in its entirety.

In sum, the network model requires both that laws remain connected in a system as coherent, economical and convenient as possible, and that the majority of its predicates remain applicable in connection with primary recognitions of empirical similarity and difference. As Quine says, "[o]ur boat stays afloat because at each alteration we keep the bulk of it intact as a going concern. Our words continue to make passable sense because of continuity of change of theory: we warp usage gradually enough to avoid rupture", or at least ruptures concerning the entire skeleton.[3]

From an empiricist point of view, the existence of such a system of laws and predicates is neither a certainty which can be guaranteed a priori, by means of some form of transcendental deduction, nor a convention that can be maintained or rejected as we like. It is something which has so far been an undeniable fact of experience, but that nothing can guarantee also for the future.

5. THE POSSIBILITY OF EMPIRICAL TESTING

The possibility of giving an answer to the problem of the empirical testability of our cognitive claims is connected to the empiricist aspect of the model just illustrated. According to the "new empiricism", "[s]cientific language is [...] a dynamic system which constantly grows by metaphorical extension", and in which alternative theoretical constructions extend metaphorically in different ways the network of functional relations between the terms of the language.[4] In principle, this is totally compatible with the persistence of common points and areas of intersection capable of allowing an empirical comparison between competing hypotheses and theories. It is true that we cannot talk about abso-

---

[1] Hesse 1974: 22f.
[2] Hesse 1974: 16; see also the Conclusion (pp. 43f).
[3] Quine 1970: 4.
[4] Hesse 1974: 4f.

lutely observational sentences, whose meaning and truth value are logically determined by experience (semantical conception of observation). Nor can we talk about expressions bound to a categorial scheme fixed once and for all (weak thesis). But we cannot exclude a priori the independence of certain empirical applications of some words with respect to alternative hypotheses and theories.

The vicious circularity objection can be formulated as follows: "if the use of all observation predicates carries theoretical implications", it becomes difficult to understand how they can "be used in descriptions which are claimed to be evidence for these same theories".[1] The network model allows us to break the "vicious circularity of truth and meaning, for at any given time *some* observation statements result from correctly applying observation terms to empirical situations according to learned precedents and independently of theories, although the relation of observation and theory is a self-correcting process in which it is not possible to know at the time which of the set of observation statements are to be retained as correct in this sense, because subsequent observations may result in rejection of some of them".[2]

In order to articulate the solution better, it is convenient to focus on two interpretations of the circularity objection.

(1) The *first formulation* states that "by 'theory-laden' is meant that the terms used in the observation report presuppose the *truth* of the very theory under test". In this case, it is trivially correct to say that such an "observation report cannot contribute evidence" for the theory under consideration. I will mention one of the most famous examples from the history of philosophy of science, also discussed by Duhem in debate with Poincaré: if in a perhaps "complex and disguised" way we *define* property *P* ('motion in a straight line with uniform speed') to be equivalent to property *Q* ('motion under no forces'), "this definition implies the truth of the law of inertia, and an observation report" which states that a body *a* endowed with property *Q* has also property *P*, does not prove anything about such a law.

In the network model, the problem cannot be set up in such simplistic terms. The law which is being tested is considered part of a network of empirical associations and nomological connections concerning, for example, the relations between measuring rods and clocks, the concept of 'rigid' rod, geometrical relationships, the properties of propagation of light studied by classical optics, and so on. Consequently, it is not possible to exclude a priori the possibility of initiating a complex process of confirmation of the considered principle "in which the correct applicability of predicates, even in observation reports, is strongly influenced by the truth of some laws into which they enter, and only vanishingly influenced by others".[3]

---

[1] Hesse 1974: 33.
[2] Hesse 1974: 43.
[3] Hesse 1974: 33f.

Therefore, we can reply to the first version of the circularity objection by saying "that although the 'meaning' of observation reports is 'theory-laden', the truth of particular theoretical statements depends on the coherence of the network of theory and its empirical input".[1] In the same spirit, Duhem pointed out to Poincaré that even principles which taken in isolation resist *direct* empirical refutation may collapse under the weight of the contradictions which experience inflicts upon the consequences of the system they belong to, considered in its totality.[2]

(2) The *second formulation* of the circularity objection connects this objection to the problem of commensurability between alternative theoretical systems. The question is the following: if the meaning of observation reports is determined, either completely or in part, by the theories referred to, how will it be possible to use the outcome of a possible crucial experiment between rival theoretical systems, since the description of this experiment "cannot have the same meaning as evidence for two different theories"?

It is possible to suggest an answer on the basis of the network model. The predicates of the 'observational reports' are tied to the acceptance of a complex of nomological connections. But this does not rule out the possibility that some laws can be more relevant than others in pragmatically fixing the correct use. Therefore, it is not possible to exclude the possibility that two theories, very different as regards the majority of their implications, share some predicates whose use is governed by the same 'hard core' of general connections. For example, "Newtonian and Einsteinian dynamics differ radically in the laws into which the predicate 'inertial motion' enters, but they share such hard-core predicates as 'acceleration of falling bodies near the earth's surface', 'velocity of light transmitted from the sun to the earth', and so on, and they share some of the laws into which these predicates enter". Therefore, it will be possible in principle to utilise the "area of *intersection* of laws" in order to carry out crucial experiments whose outcomes may be described by means of predicates with a relatively common meaning.

Of course, it would be useless to try to 'withdraw' into a mythical absolutely neutral observation language, free of any theoretical contamination, since nothing of the kind exists. Instead, what can be done, and is done, is to refer to the "area of intersection of predicates and laws between the theories".[3] The actual terms of a comparison between rival theories, or of the testing of theoretical hypotheses within a theory, cannot be specified on an abstract, general level, without considering the peculiarities of the single examined cases, in all their speci-

[1] Hesse 1974: 35.

[2] See Duhem 1962a, in particular pp. 215f.

[3] Hesse 1974: 35.

ficity and concreteness (think, for example, of Allan Franklin's recent philosophical analysis of the experiment on parity conservation).[1]

5 1. EMPIRICAL TESTING AND KUHN'S EXEMPLARS

I would like to further illustrate the question of the relation between primitive recognitions of similarity and the possibility of empirical testing by briefly considering the relations between the network model and one of the main arguments deployed by Kuhn to explain the incommensurability connected with scientific revolutions.

Among the concepts which form Kuhn's notion of 'paradigm' (understood as 'disciplinary matrix'), there is that of 'paradigmatic example' or 'exemplar'. This concept has a central importance, which has been adequately appreciated only in recent years.[2] Kuhn clarifies the nature of exemplars using Wittgenstein's ideas of family resemblance and language game. Exemplars are "*concrete problems with their solutions*",[3] which, in the exposition and in the teaching of a theory or a discipline constitute models to be followed for the solution of the remaining problems of 'normal science'. Their main characteristic is that they perform the same function as general explicit rules, whilst maintaining themselves on the level of particular and concrete exemplification.

There is a precise parallel between the concept of a paradigm and the network model, because Kuhn too wanted to show how the formation and the learning of meanings inextricably involve, as well as the capacity of using explicit definitional criteria which can be formulated linguistically, also the disposition to tacitly pick out networks of similarities and dissimilarities between empirical situations. Reference to exemplars should illustrate the way language 'adheres' to experience, thus clarifying the ways in which it is possible to learn the language and the theories – "learning language and nature together by ostension".

Kuhn says: "[o]ne of my claims is [...] that we have too long ignored the manner in which knowledge of nature can be tacitly embodied in whole experiences without intervening abstraction of criteria or generalisations. [...] One of the things upon which the practice of normal science depends is a learned ability to group objects and situations into similarity classes which are primitive in the sense that the grouping is done without an answer to the question, 'similar with respect to what?' One aspect of every revolution is, then, that some of the similarity relations change. Objects which were grouped in the same set before are grouped in different sets afterwards and *vice versa*. Think of the sun, moon, Mars, and earth before and after Copernicus; of free fall, pendular, and planetary motion before and after Galileo; or of salts, alloys, and

---

[1] Franklin 1986.
[2] See Lanfredini 1988, ch. 3; Hoyningen-Huene 1993
[3] Kuhn 1977c: 306 (italics added).

a sulphur-iron filing mix before and after Dalton. Since most objects within even the altered sets continue to be grouped together, the names of the sets are generally preserved. Nevertheless, the transfer of a subset can crucially affect the network of interrelations among sets. [...] When such a redistribution of objects among similarity sets occurs, two men whose discourse had proceeded for some time with apparently full understanding may suddenly find themselves responding to the same stimulus with incompatible descriptions or generalizations. Just because neither can then say, 'I use the word element (or mixture, or planet, or unconstrained motion) in ways governed by such and such criteria', the source of the breakdown in their communication may be extraordinarily difficult to isolate and by-pass".[1]

Kuhn uses the analysis of the function of exemplars in order to show how, as a consequence of important scientific changes, 'breakdowns' may occur in the communication between researchers adhering to different research paradigms. This is certainly a possible outcome of scientific revolutions, and it cannot be ignored.

But, considering the question from the point of view of the network model, we can understand how some primitive groupings of similarity, embodied in certain paradigmatic examples, *may* remain unchanged even through major changes, thus providing a relatively neutral basis of empirical comparison between alternative theories and hypotheses. We can prove a priori neither that there will always be total empirical incomparability, nor that there will always remain some areas of intersection at the level of verbal connections and primary recognitions of similarity. What *is* proved, instead, is that one or the other case may possibly occur: actual discussion between scientists, or historical reconstruction will yield a certain interpretation of what is going on or of what has happened.

For those who, like myself, want to retain a good dose of empiricism, it is pleasant to note that *there are no philosophical conceptions powerful enough to ensure a priori or to exclude a priori the empirical testability of our cognitive claims*. It all depends upon what we say, how we say it, and upon our ability in planning and carrying out the appropriate experiments – as all good experimenters have always known.

[1] Kuhn 1987b: 275f.

Chapter IV

LOGICAL IDEALISM, FORMAL COHERENCE, AND MATERIAL CORRESPONDENCE

1. THE CRITICAL CONCEPTION OF TRUTH: THE PRIMACY OF THE JUDGEMENT ON THE OBJECT

1.1. KANT'S FORMAL IDEALISM

In an essay of the early sixties, Feyerabend presented his point of view as a resumption of certain aspects of Kant's philosophy. He rejected the idea of absolutely valid transcendental conditions of knowledge, but retained that of all-pervasive theoretical assumptions as ways of looking at the world: "what is regarded as 'nature' at a particular time is our own product in the sense that all features ascribed to it have first been invented by us and then used for bringing order into our surroundings".[1] This idea is partly incorporated in the network model which I just introduced, since in the network model, while our cognitive claims are empirically testable, a considerable role is set aside for linguistic, theoretical and methodological-axiological epistemic conditions.

It remains for me now to formulate a notion of objectivity which secures, along with the empiricist component of positive philosophy which I argued for in the previous chapter, the objectivist component of positive philosophy.

But, before I set forth my proposal, in which I will try to harmonise the somewhat contrasting aspects revealed by the analysis of the cognitive process, I must highlight other relevant data. In order to do this, I will consider some epistemological conceptions alternative to the one that I want to suggest. Having already secured a good dose of empiricism, I will start with a moderate version of idealism. I will complete the preliminary part of the argument in the next chapter, assessing the chances of metaphysical realism and discussing the contrast between scientific realism and instrumentalism.

The conception I am about to examine can be considered a moderate form of idealism, since it tries to reconcile rationalist and idealist elements with empiricist and realist ones. I will construct this general theoretical position, starting from Kant's 'formal idealism' and considering some adaptations of it intended to make it compatible with the scientific and philosophical developments of this century.

However, my 'labelling' is not absolute. My classification of the line of thought I am about to sketch as idealist depends only on the fact that my positive point of view so classifies it. I am fully aware – and I will not fail to emphasise it – that many exponents of idealism have re-

[1] Feyerabend 1981c: 45.

peatedly presented their ideas as an attempt to overcome the two contrasting metaphysics of idealism and realism. Except in extreme cases, only a relative value can be attributed to such classifications. Likewise, I suspect that a metaphysical realist would not hesitate to consider even my positive philosophy as a variant of idealism, despite the fact that I do not see it as being so in the least. *I consider it rather a non-idealist rejection of metaphysical realism, or, better, a non-idealist rejection of the metaphysical interpretation of scientific realism.*

In order to characterise formal idealism, I will refer to the theory of truth which can be extrapolated from the *Critique of Pure Reason*. This choice is due to three main reasons.

(1) The first reason depends upon requirements of presentation. I am convinced that the theory of truth allows the individuation of a 'minimal' nucleus of transcendental epistemology which is relatively sheltered from the many and fundamental interpretative disagreements which characterise Kantian studies. This should assist me in producing an account which is both systematically unitary and historically not too arbitrary.

The other two reasons are specifically theoretical.

(2) In the first place, there is a substantial affinity between the general inspiration of my work and the alethic approach to formal idealism.[1]

In recent years, it has been repeatedly noted that Strawson's famous interpretation of the *Critique of Pure Reason* in *The Bounds of Sense* (1966) is strongly conditioned by the "underlying assumption [...] that Kant's argument in the 'Analytic' [...] concerns [...] only the sufficient conditions for giving *sense* to the *concept* of the objective or the *general contrast* between subject and object". However, for some more recent interpreters, Kant's argument concerns, before any other thing, "the conditions for actually confirming *particular judgements* or claims to empirical knowledge".[2] My research agrees with this interpretation: it starts from the problem of the epistemic justification of our cognitive claims and of the various presuppositions which lie at their basis.

(3) The Critical conception of truth presents itself as a mixture of the coherentism and correspondentism which characterises the network model. Consequently, the Critical conception of truth can be considered a possible general philosophical framework for the theses defended in the previous two chapters about the relations between language, theory and experience. Therefore, individuating the difficulties of formal idealism from the perspective of the theory of truth should constitute an adequate preamble for the position that I want to argue for.

---

[1] My discussion of Kant's conception of truth is largely a re-elaboration of Parrini 1994a.

[2] Guyer 1987: 231; see also Parrini 1994a: 227f, n. 34 (the reference is to Strawson 1966).

A. Kant's theory of truth

In Sections III and IV of the Introduction to the Second Part ('Transcendental Logic') of the 'Transcendental Doctrine of Elements',[1] Kant introduces his conception of truth by making some preliminary distinctions between 'nominal *definition*' and 'general and sure *criterion* of truth of any and every knowledge'; between 'general *material* criterion' and 'general *formal* criterion'; and, lastly, between '*negative* general formal criterion' and '*positive* general formal criterion'. With an attitude which in some ways prefigures the epistemological 'neutrality' of Tarski's definition of truth for formalised languages,[2] Kant does not attribute epistemological relevance to the definitional aspect of the question. He believes it to be somehow solved by the Thomist-Scholastic formula (derived from Aristotle) according to which *veritas est adaequatio intellectus et rei*. Kant says that he takes for granted the nominal definition of truth, on the basis of which truth "is the agreement of knowledge with its object".[3]

Kant does not concentrate on the *definitional* aspect of truth: he concentrates instead on the epistemological aspect, which involves the specification of a general character sufficient to distinguish true cognitive claims from false ones. Kant rejects the possibility of a general material criterion, which would allow the truth of any judgement to be established without taking into account its specific assertive content. Instead, he vindicates the possibility of identifying a general *formal* criterion. In the case of a priori judgements (synthetic and analytic), this formal criterion is a necessary and sufficient condition of the truth of the judgement, and it is therefore a *positive* criterion. In the case of synthetic a posteriori judgements, it constitutes a necessary but not sufficient condition, and is therefore purely *negative*.[4]

The general *formal* (positive and negative) criterion of truth is given, in the first place, by the principles of general logic, starting with the principle of non-contradiction (the "highest principle of all analytic judgments",[5] and with the 'Principle of categorical inferences of reason': "[w]hat belongs to the mark of a thing belongs also to the thing itself; and what contradicts the mark of a thing contradicts also the thing itself" ("*nota notae est nota rei ipsius: repugnans notae, repugnat rei ipsi*").[6] Since logic spells out the "universal and necessary rules of the understanding", along with the formulation of such rules, logic supplies also the "criteria of truth". The judgements that contradict these rules must be false, "[f]or the understanding would thereby be made to contradict its own general rules of thought, and so to contradict itself."

[1] Kant 1985: 97-101 [A57-64 = B82-8].
[2] See Tarski 1964: 33f.
[3] Kant 1985: 97 [A58 = B82].
[4] Kant 1985: 97 [A57f = B82f].
[5] Kant 1985: 189-91 [A150-3 = B189-93].
[6] Kant 1992. 617f [9. 123]; see also Introduction, sec. VIII (pp. 564ff [9: 58ff])

According to Kant, however, these criteria "concern only the form of truth, that is, of thought in general; and in so far they are quite correct, but are not by themselves sufficient. For although our knowledge may be in complete accordance with logical demands, that is, may not contradict itself, it is still possible that it may be in contradiction with its object. The purely logical criterion of truth, namely, the agreement of knowledge with the general and formal laws of the understanding and reason, is a *conditio sine qua non*, and is therefore the negative condition of all truth".[1] The "purely logical criterion of truth" can function also as a positive condition only in the case of analytic judgements (affirmative or negative), because their "truth can always be adequately known in accordance with the principle of contradiction. The reverse of that which as concept is contained and is thought in the knowledge of the object, is always rightly denied. But since the opposite of the concept would contradict the object, the concept itself must necessarily be affirmed of it".[2]

The formal criterion of truth is not only made up of the *logical-analytical conditions* concerning the non-contradictory character of the concepts and thoughts contained in judgements. If we take into account the whole framework of transcendental epistemology, it is clear that it consists also in the conformity of our cognitive claims to the synthetic a priori forms of knowledge – that is, to the *logical-transcendental conditions* concerning the possibility of the objects of experience.

This possibility is circumscribed, on close inspection, by the conditions of understanding without which no object can in any way be thought of, as much as by the conditions of sensibility without which no object can be intuited. Therefore, the formal criteria include all subjective, a priori structures of knowledge, namely: the forms of space and time; the categories; the schemes as mediating elements between understanding and sensibility; and finally the principles of pure understanding, especially the causality principle and the other two analogies of experience. In a way which I will not be able to discuss here, the methodological rules of reason concerning the unity of nature and the systematic order of natural laws, can also be included in the formal criterion of truth – although they are given a role which is simply regulative, and not constitutive.[3]

All these conditions can be considered, for different reasons, as part of the presuppositional tools with which the creators of modern science – from Galileo to Newton, from Torricelli to Stahl – have interrogated nature, not randomly, but by forcing her to answer questions conceived according to the plan of reason.[4]

---

[1] Kant 1985: 98 [A59f = B84].

[2] Kant 1985: 190 [A151 = B190f]

[3] See the analysis of the methodological rules of reason in particular in the "Transcendental Dialectic" and in the *Critique of Judgement*.

[4] See Kant 1985: 20 [Bxiif]; the connection between a priori presuppositions and natural science is clearly formulated also in Kant 1993: 149 [22: 507].

B. Kant's epistemic conditions as conditions of the possibility of the objects of experience

The distinction between logical-analytic conditions and constitutive logical-transcendental conditions is obviously related to the well-known Kantian dichotomy between analytic and synthetic a priori judgements. If we consider neither the supposed apodictic character of synthetic a priori presuppositions, nor Kant's identification of such presuppositions with certain specific assumptions of the knowledge of his time (such as, for example, Euclideanism and the causality principle), this distinction can be made to correspond with the distinction between linguistic rules and theoretical assumptions, which I talked about in Chapter II when I distinguished between the different dimensions of epistemic relativism. The methodological principles correspond on the other hand to the methodological and axiological assumptions. I will therefore follow Allison in calling all these presuppositions 'epistemic conditions' to distinguish them from logical, psychological and ontological conditions.[1]

The characteristic aspect of the transcendental theory of truth and objectivity is the fact that the constitutive epistemic conditions, despite their subjective nature, are not merely conditions of the possibility of experience, i.e., of the objects' being perceived and thought, and therefore *known*; they are as well conditions of the possibility of the *objects* of experience themselves. This is explicitly said in the Section on 'The Highest Principle of all Synthetic Judgements',[2] and is obviously connected with the distinction between phenomenal objects and things in themselves which, unlike the phenomena, do not fall under the jurisdiction of our forms of representation.

The epistemic conditions "contain in themselves the ground of the possibility of experience viewed as the sum of all knowledge wherein objects can be given to us". They are "true *a priori*" because it is they who constitute "the source of all truth (that is, of the agreement of our knowledge with objects)".[3] No cognitive claim and no judgement can contradict them "without at once losing all content, that is, all relation to any object, and therefore all truth".[4]

In this conception, attributions of reality and unreality, existence and non-existence, are connected not only with the material, sensible conditions of experience, but also with the subjective constitutive conditions. As well as in 'The Postulates of Empirical Thought in General' (concerning the modal categories of possibility, existence and necessity),

---

[1] Allison 1983: 10ff, 25ff; 336, n. 30 for the equivalence between epistemic *forms* and *conditions* of knowledge.

[2] Kant 1985: 191-4 [A154-8 = B193-7]

[3] Kant 1985: 258 [A237 = B296].

[4] Kant 1985: 100 [A62f = B87]. This relationship between a priori truth and principles of understanding is analysed, in inverted form, also in the Section 'The Highest Principle of All Synthetic Judgments' (Kant 1985: 193f [A157f = B196f]).

this idea is formulated in 'Note 3' of the 'Refutation of Idealism', added in the second edition of the *Critique*. After having proved that "the existence of outer things is required for the possibility of a determinate consciousness of the self", Kant observes that from such a thesis it "does not follow that every intuitive representation of outer things involves the existence of these things, for their representation can very well be the product merely of the imagination (as in dreams and delusions)". According to Kant, one can ascertain "[w]hether this or that supposed experience" is or is not "purely imaginary" only "from its special determinations, and through its congruence with the criteria of all real experience", that is, with those which I called constitutive logical-transcendental conditions of truth.[1] As Kant already stated in 'Remark III' of the *Prolegomena*, "the difference between truth and dreaming is not ascertained by the nature of the representations, which are referred to objects (for they are the same in both cases), but by their connection according to those rules, which determine the coherence of the representations in the concept of an object, and by ascertaining whether they can subsist together in experience or not".[2]

As Kant says particularly clearly in the famous letter to Marcus Herz of February 21st, 1772, this conception of truth and reality is the correlate of a theory of knowledge which conceives a cognisant understanding neither as an *intellectus ectypus*, whose representations are the mere passive effect of the action exerted by things on our spirit, nor as an *intellectus archetypus*, similar to the divine intellect, in which the representation itself produces the object corresponding to it. Rather, the human mind is a mixture of passivity and spontaneity: knowledge aims to unify a manifold of sensible data (acquired passively through certain forms of sensibility) by means of certain a priori concepts, schemes and principles, which are the product of intellectual spontaneity.

Therefore, it is no longer possible to rely upon a metaphysics, or an ontology, conceived as science of being qua being, where being is an object which transcends the domain of all possible experience. Consequently, metaphysics can survive only if redefined as critical investigation of those a priori "conditions of the *possibility of experience*", which are at the same time "conditions of the *possibility of the objects of experience*":[3] "the proud name of an Ontology [...] must [...] give place to the modest title of a mere Analytic of pure understanding".[4]

C. *Kant's conception of truth and his formal idealism*

The arguments with which Kant tries to warrant these conclusions in the *Critique* are the transcendental deduction of the categories and the demonstrations of the principles. Despite the fact that these arguments

---

[1] Kant 1985: 247 [B278f].
[2] Kant 1995:34 [290].
[3] Kant 1985: 194 [A158 = B197].
[4] Kant 1985: 264 [A247 = B303].

still constitute some of the most enigmatic and controversial pages of the entire philosophical literature, there is at least one passage which can be considered relatively non-problematic. It is the passage where, analysing what is to be understood by object of representations, and therefore by object of knowledge, Kant declares that the Critical-transcendental conception is the only one capable of making it into something other than an elusive chimaera. "[A]ppearances are themselves nothing but sensible representations, which, as such and in themselves, must not be taken as objects capable of existing outside our power of representation. What, then, is to be understood when we speak of an object corresponding to, and consequently also distinct from, our knowledge? It is easily seen that this object must be thought only as something in general = *x*, since outside our knowledge we have nothing which we could set over against this knowledge as corresponding to it". "Now we find – Kant continues – that our thought of the relation of all knowledge to its object carries with it an element of necessity; the object is viewed as that which prevents our modes of knowledge from being haphazard or arbitrary, and which determines them *a priori* in some definite fashion. For in so far as they are to relate to an object, they must necessarily agree with one another, that is, must possess that unity which constitutes the concept of an object.

But it is clear that, since we have to deal only with the manifold of our representations, and since that *x* (the object) which corresponds to them is nothing to us – being, as it is, something that has to be distinct from all our representations – the unity which the object makes necessary can be nothing else than the formal unity of consciousness in the synthesis of the manifold of representations".[1]

Therefore, the conception of truth hinging on the dichotomy between formal and material conditions can be considered as the obvious alethic correlate of the well-known Kantian idea of knowledge, or of the known object, as synthesis of form and matter. More precisely, the theory of truth – leaving aside very controversial issues, such as, for example, the ontological value of transcendental epistemology or the justification of the causality principle – limits itself to condensing those aspects of Kant's thought by which it can be defined as an idealism of the formal conditions of knowledge. Kant himself called it Critical or *formal* idealism, in order to distinguish it from Descartes' problematic idealism and Berkeley's dogmatic idealism.[2] These aspects of Kantianism are emphasised also in recent work by Pippin, Allison and Guyer.[3] Most significantly, they constituted the vital core of the Marburg school's, and, still more, of Cassirer's interpretation of the *Critique* and of the *Prolegomena*.

---

[1] Kant 1985: 134f [A104f].

[2] See the concluding part of Remark III of the *Prolegomena* (Kant 1995: 36f [293f]), as well as the Appendix on Garve and Feder's review of the *Critique* (Kant 1995: 112ff [372ff]).

[3] Pippin 1982; Allison 1983; Guyer 1987.

Today, after the works of Paulsen, Heidegger, Vleeschauwer, Heimsoeth, Martin, etc., which have shed light on the metaphysical, ontological and anthropological facets of the 'transcendental turn', we are generally persuaded of the unilateral character of such a reading, if we understand it as an overall interpretation of Kant's thought. Yet, it continues to exert a strong and undeniable fascination because of its capacity to focus the formal or logical (as Cassirer will say) nature of transcendental idealism. Placing this aspect of the Critical -transcendental turn to the forefront means, firstly, to make its revolutionary import consist in the new perspective it provides on the relationship between subject and object of knowledge. This helps us distinguish Kant's position from Berkeley's idealism, and allows us to fully appreciate Kant's aspiration to address and solve the epistemological problem without committing himself to ontological realism, idealism, phenomenalism or materialism.

According to Cassirer, before Kant "the thing and the self had always been projected, in order to be understood in their connection, upon a common metaphysical background". Kant sought "the logical, universally valid and fundamental form of experience in general, which must be binding in the same way both for 'internal' and for 'external' experience". In this sense, we may apply to Kant's philosophy "the words of Schiller's famous epigram: it knows nothing about the thing and nothing about the soul". The object of Kant's inquiry is "no longer composed of the things, but of the judgments on the things".[1] For Kant, "judgment and object are strictly correlative concepts, so that in the critical sense, the truth [= reality] of the object is always to be grasped and substantiated only through the truth of the judgment".[2] In Critical-transcendental philosophy, "it is not because there is a world of objects that there is for us, as their impression and image, a world of cognitions and truths". On the contrary: if there is for us an organisation which we can designate as an order not merely of "impressions and representation", but also of objects, this is because there are "unconditionally certain judgments – judgments whose validity is dependent neither on the individual empirical subject from which they are formed nor on the particular empirical and temporal conditions under which they are formed".[3]

As Abel Rey said, illustrating the passage from the old positivism to Poincaré's new positivism, it was essential to realise that "[o]bjective experience and mind are functions of each other, imply each other, and exist by virtue of each other".[4] Through Dummett's mediation, something similar has come about in American philosophy. With Putnam's passage from external to internal realism, at Harvard, the idea "that objects and reference arise out of discourse rather than being

[1] Cassirer 1922· 662.
[2] Cassirer 1981: 285.
[3] Cassirer 1981. 148.
[4] Rey 1907. 392f, cited in Frank 1975: 9; see ch I, sec. 2, above

prior to discourse" is being taken into serious consideration again: "[r]eference [...] is not something prior to truth; rather, knowing the conditions under which sentences about, say, tables, are true is knowing what 'table' refers to (as on a disquotational theory of *reference*)".[1]

1.2. CASSIRER'S LIBERALISATION OF KANTIANISM

A. The conception of truth and objectivity

The developments of scientific investigation have affected the presupposition on which Kant's conception was built, namely the universal and necessary validity of certain principles of judgement. What can be preserved then of Kant's remarkable conception of objectivity and truth? I will give my answer in Chapter VI. For the present, I want to point out the fact that not all those who endorse the crisis of the specific presuppositional apparatus defended in the *Critique* have, because of this, declared themselves to be anti-Kantians. Many philosophers have defended transcendental epistemology, reckoning that in order to do so they need only show the unsustainability of rigidly empiricist conceptions of concept formation and the largely inventive and constructive character of scientific activity. In order to display the merits and the limits of this outlook, I will consider two significant twentieth century versions of criticism, namely, Cassirer's (this section, 1.2.) and Preti's (section 2.1., below).

I will start from the neo-Kantian philosopher who more than anyone else experienced the crisis of transcendental epistemology brought about by the new scientific theories. Cassirer started his work of adaptation of Kantianism in 1910, with *Substanzbegriff und Funktionsbegriff*. Although quantum and relativistic physics had not yet been fully developed, Cassirer realised immediately the 'subversive' philosophical potential of the emerging ideas. He took action straight away by developing a sort of 'liberalisation' of Kant's rigid formulations. This liberalisation made very able use of the anti-empiricist aspects of the epistemological views of Hertz, Poincaré and Duhem. It was especially directed against Mach's conception, interpreted as a form of radical sensationalism and phenomenalism.

It will not escape a careful reader that Cassirer sometimes conflates the notions of relativity and relationality, and that his use of the results of the mathematical, physical, physiological and psychological sciences lacks the necessary philosophical mediations (I think this is particularly clear in his use of Helmholtz's contributions). But this does not impair the substance of his epistemological construction which, in the wake of Kant's analysis of the object of representation, culminates in a characterisation of scientific truth and objectivity which is to a certain extent independent of the assumption of transcendental conditions specified once and for all.

---

[1] Putnam 1983a: xvi

According to Cassirer, the correspondence view of truth – connected with the metaphysical realist postulation of a domain of objects which transcend the cognitive process, and to which knowledge should somehow conform in order to be true – is a dogmatic conception inevitably leading to scepticism. We must accept logical or formal idealism if we want to avoid the sceptical outcome. Logical idealism, as it is professed in the essay of 1920 on relativity "does not measure the truth of fundamental cognitions by transcendent objects, but it grounds conversely the meaning of the concept of the object on the meaning of the concept of truth".[1] Cassirer had specified that meaning already in *Substanzbegriff und Funktionsbegriff* (1910), trying to ascribe to it the changeable, somehow conventional, character of the human mind's spontaneous contribution to knowledge.

In conformity with the methodological aspects of Kant's thought, connected to the regulative principles of uniformity and systematicity, Cassirer gave the notion of truth the meaning it has in the actual practice of scientific investigation: "[s]cience has, and can have, no higher criterion of truth than unity and completeness in the systematic construction of experience as a whole. Every other conception of the object lies outside its field; it must 'transcend' itself even to be able to conceive the problem of another sort of objectivity".[2] What binds the autonomous activity of the intellect in the construction of scientific syntheses is neither an unattainable metaphysically transcendent reality, nor an *isolated* perception; it is only the "*system* of perceptions in their order and connection. It is true that this order is never to be established in a single system of concepts, which excludes any choice, but it always leaves room for different possibilities of exposition; in so far as our intellectual construction is extended and takes up new elements into itself, it appears that it does not proceed according to caprice, but follows a certain law of progress. This law is the ultimate criterion of 'objectivity'; for it shows us that the world-system of physics more and more excludes all the accidents of judgement, such as seem unavoidable from the standpoint of the individual observer, and discovers in their place that necessity that is universally the kernel of the concept of the object".[3]

It is clear why I consider the Kantian and neo-Kantian position a moderate form of idealism. Their aim is a conception of the object which somehow overcomes the traditional opposition between realism and idealism, along the lines traced by Kant in the *Critique*. After having distinguished between the empirical and the transcendental meaning of the couple appearance / reality, Kant defined himself as empirical realist and transcendental idealist: empirical realist because he asserted the transcendence of the object of knowledge with respect to those particular empirical objects which are the concrete individual subjects; transcendental idealist because he maintained the immanence of that same

[1] Cassirer 1953b: 391.
[2] Cassirer 1953a: 187, translation modified.
[3] Cassirer 1953a: 187.

object with respect to the logical-transcendental presuppositions of the cognitive process.[1]

Cassirer draws on this idea also in trying to overcome the opposition between absolutism and relativism: "[i]f we determine the object not as an absolute substance beyond all knowledge, but as the object shaped in progressing experience", then "this object may be called 'transcendent' from the standpoint of a psychological individual" although "from the standpoint of logic and its supreme principles [...] it is to be characterized as purely 'immanent' [...] This simple thought alone constitutes the kernel of critical 'idealism'. [...] The ideality, which is here alone asserted, has nothing in common with the subjective 'presentation'; it concerns merely the objective validity of certain axioms and norms of scientific knowledge. The *truth* of the object – this alone is affirmed – depends on the truth of these axioms, and has no other and no firmer basis. It is true that there results, strictly speaking, no absolute, but only a relative being. But this relativity obviously does not mean physical dependency on particular thinking subjects, but logical dependency on the content of certain universal principles of knowledge. The proposition, that being is a 'product' of thought, thus contains no reference to any physical or metaphysical causal relation, but signifies merely a purely functional relation, a relation of superordination and subordination in the validity of certain judgements".[2]

B. A cumulativist view of scientific change

This way of adapting formal idealism's conception of scientific objectivity to the developments of science is connected with Cassirer's substantially 'continuist' and 'cumulativist' view of scientific, observational and theoretical changes. There are, in this view, "certain *material* features of the system, that are unaffected by the transition from one stage to the following: the variation does not absolutely cancel the earlier stage, but allows it to remain in a new interpretation. All the observations of Tycho de Brahe enter into the system of Kepler, although they are connected and conceived in a new way".[3]

As is stated in the essay on relativity, the new interpretative connections depend upon the development of new 'conceptual tools'. Because of this, the history of physics cannot be reduced to the chronicle "of the discovery of a simple series of 'facts'"; it is also the history of new unifying ideas. But "the discovery of ever new and more special means of thought" should not be seen as a 'rupture' in the continuity of the process of cognitive unification: "in all change of these means of thought there is nevertheless revealed, as surely as physics follows the 'sure course of a science', the unity of those methodic principles upon which rests the formulation of its question".[4]

---

[1] See, for example, Allison 1983.
[2] Cassirer 1953a: 297f.
[3] Cassirer 1953a: 321.
[4] Cassirer 1953b. 421.

As was already declared in the work of 1910, "the changing phases of the concept of experience do not lie absolutely outside of each other, but are connected by logical relations. The system and the convergence of the series takes the place of an external standard of reality. [...] The one reality can only be indicated and defined as the ideal limit of the many changing theories; yet the assumption of this limit is not arbitrary, but inevitable, since only by it is the continuity of experience established. No single astronomical system, the Copernican as little as the Ptolemaic, can be taken as the expression of the 'true' cosmic order, but only the whole of these systems as they unfold continuously according to a definite connection".[1]

2. TRANSCENDENTAL CRITICISM, SCEPTICISM, AND METAPHYSICAL REALISM

2.1. THE CONNECTION BETWEEN SCEPTICISM AND METAPHYSICAL REALISM

Cassirer's reformulation of formal idealism takes into account the changeable and contingently conventional aspects of scientific development. It rests upon two fundamental points:

(1) the Kantian notion of cognitive objectivity and truth, reinterpreted as "the ideal limit of the many changing theories";[2]

(2) the interpretation of the transcendentally a priori principles of knowledge not as already specified presuppositions at the basis of our conceptual organisation of experience, but as "ultimate logical invariants" of experience, which confer unity on the cognitive process, constituting the immanent law of its development.[3]

In more recent years, Giulio Preti, in Italy, has further developed this line of thought. His proposal aspires to connect in a non trivially eclectic synthesis demands which came from the most significant philosophical experiences of our time: neo-Kantianism and phenomenology, conventionalism and neopositivism, historicism and pragmatism. His integration was not always respectful of the motivations at the bottom of the different traditions of thought that he brought together. Later on, I will show that the typically neopositivist verificationist approach to the problem of knowledge cannot be superimposed on the neo-Kantian and phenomenological approach, based upon the appraisal of scepticism. I will discuss Preti's attempt in order to introduce the principal difference between these two formulations, without losing sight of their similarity. In Chapter VI, this will allow me to outline the conception of truth and

[1] Cassirer 1953a: 321f
[2] Cassirer 1953a. 322.
[3] Cassirer 1953a: 269.

of objectivity required by the appreciation of the rationalist component of positive philosophy.[1]

1. The neo-Kantian and phenomenological aspect of Preti's thought. The overcoming of the metaphysical formulation of the epistemological problem – I find the basic inspiration of Preti's conception of cognitive objectivity closer to the ideas which emerged later in the new philosophy of science, than to the neoempiricist conception of theories, strongly conditioned by the verifiability principle. The object of knowledge is conceived as the rational re-elaboration of the empirical material on the basis of the historically changeable forms and principles regulating cognitive activity. In Husserl's rather than in Kant's terms, the object of knowledge is conceived as the noetic / noematic synthesis of intentional forms and hyletic contents. Therefore, there is an undeniable agreement between the idea that theoretical concepts are intentional forms, which 'animate' the hyletic contents of perception, making them "open to other contents and signifying of them",[2] and Feyerabend's overturning of the relation between theory and experience, by which the meaning of the observation sentences 'trickles down' from the theory to empirical observation, rather than vice versa.

In agreement with one of the most marked tendencies of neo-empiricist epistemology, Preti tried to trace back the interpretative structures to the linguistic dimension of scientific discourse. But this took place in the context of a general view of meaning and of the transcendental subject, which grants to these structures a role analogous to that which Kuhn assigns to the conceptual components of the disciplinary matrix.

Preti referred to Bruno Bauch in order to characterise the transcendental subject not as "an ego in general self-referred and self-intuited",[3] or as "a constellation of fixed stars of thought",[4] but as "the formal structure of scientific discourse (granting, as a convenient hypothesis, that it be unique)".[5] The transcendental subject is the locus of a "relative historical a priori", largely conventional, constituted by conceptual structures comparable with Husserl's noetic forms and with Dewey's meanings. These structures can be found in the process of the formation and transformation of sciences, within which they determine "the sphere of relations to which experiences must be referred, and by means of which experiences can be interpreted and evaluated".[6]

Thus conceived, the transcendental subject no longer aspires to constitute "a *foundation*, nor, for its part, does it require a founda-

---

[1] For the following reconstruction of Preti's epistemological thought, and for the relevant bibliographical information, see Parrini 1988a.

[2] Preti 1976f: 504

[3] Preti 1976e: 465

[4] Preti 1976c: 415.

[5] Preti 1957: 130.

[6] Preti 1976c: 416f

tion".[1] It can be seen as a dynamic centre of meaningfulness, where, with Dewey, we consider meaning not as "a token", constituting the static summary "of essential or common notes" codifying the past empirical data, but as a "coin", which allows the "circulation", the "exchange" and the "connection" of "far apart and heterogeneous" experiences, including future ones.[2] Science produces knowledge not because it accomplishes the impossible task of mirroring a reality in itself transcendent with respect to it, but because it elaborates a "network of symbols and concepts" which allow "a systematic interpretation of the factual contents", that is "the construction of a picture of the world, of a 'nature'".[3]

The existence of a 'real world' can be considered "a *postulate* of science" only if understood "as a *regulative idea of science itself*": "[i]t is [...] not a concept, but an idea, in the Kantian sense: it is something that is never given, and antinomies are generated as soon as it is considered as given. It is a 'truth', if it can be so called, which does not enter as an axiom (even less as a theorem or as a factual sentence) in any scientific discourse; it is an hypothesis of which, to be strictly rigorous, no science is in need. Only the requirement of the unity of science demands the possibility of a unitary picture of the 'real world' – but it is a *requirement*, not a *truth*. Accordingly, the existence of a 'real world', in itself, and therefore transcendent (in Husserl's language, as absolute and total noema, correlate of the transcendental system of noetic structures), can be considered at best as an 'ideal of reason', which has a merely *regulative* and not a *constitutive* function within knowledge, i.e., as something that assigns a task and points in a direction, *not* as something which we can consider already actualised".[4]

For Preti, this does not mean acceptance of idealism. Rather, his goal is an overcoming of idealism and of realism, considered as unwarranted opposite ontological hypostatisations of the two internal poles – the subject and the object – in which cognitive activity 'dialectises itself'. Metaphysical dualism must be overcome, giving a 'transactional' value to the 'tension' between the subjective and the objective pole, and restoring the pragmatic value of the notion of reality. Preti has repeatedly insisted on the validity of the common sense realism defended by Moore, when it is considered as the expression not of a philosophical thesis, but of the pragmatic, non-thematised, acceptance of the existence

1 Preti 1957: 132.

2 Preti 1976c: 414.

3 Preti 1968: 17. Preti (1957: 120) underlines also the anti-metaphysical import of the substitution of the notion of truth by the notion of warranted assertibility – a substitution already suggested in late Scholastic philosophy.

4 Preti 1957: 120f The desire for an ontologically 'neutral' construal of the problem of knowledge is also present in Carnap's *Aufbau* (see ch. V, below). And, in the margin of the letter in which Reichenbach told Carnap that it was impossible to do without the "Axiom des Realismus", Carnap noted in shorthand: "*In* der Wissenschaft kann man aber das Realitätsaxiom nicht aussprechen" ["*In* science, however, the axiom of realism cannot be formulated"]· see Parrini 1993: 380f n. 8.

of reality as the intentional horizon of our *Lebenswelt* and of our vital acts.[1]

2. The neopositivist aspect of Preti's thought: the meaninglessness of metaphysical realism – Despite this agreement with the key themes of the Kantian and phenomenological view, Preti presented his ideas as the expression of an "empiricist point of view" imbued with the teachings of neopositivist epistemology.[2] He thought it was possible to channel in the same direction the overcoming of the metaphysical formulation of the epistemological problem pursued by the Kantian tradition, and the implications of the verifiability theory of meaning. Once we have recognised that it is possible to avoid scepticism only by liberating the problem of knowledge from undue dogmatic metaphysical presuppositions, the task of epistemology will become that of analysing the cognitive processes according to the immanent criteria and ways of their constitution and development. But we get the same result when, on the basis of the verifiability principle, we reject metaphysical accounts of the cognitive act as senseless. In this case too, epistemology will become a positive analysis, aiming at the clarification of "the phenomenological and logical problems concerning the relationships which hold between the various forms of phenomenology of knowledge".[3]

In both conceptions, knowledge ceases to be a *problem*, in order to become a *fact* which can be analysed. Therefore, it will be possible to assert that a coherently developed logical idealism turns into positivism and vice versa: "[t]he new positivism, critical rationalism which situates itself in the point where idealism and positivism meet in their programmatic and methodological requirement, is therefore anti-metaphysical. Its task is the investigation of the forms and structures of experience and of culture; its fundamental postulate is the principle of immanence; the concepts which it yields are not hypotheses about the reality of things, but methods for the resolution of experience on the plane of intellectual and rational thought".[4] In this positive guise, epistemology can retrieve even Kant's idea of a Critical ontology, different from the dogmatic metaphysics rejected in the Transcendental Dialectic: its task is the "analysis of the structures of the experience which posits the object of knowing". "Critique of knowledge" becomes "Critical ontology", in so far as it is an analytic reconstruction of the "transcendental notions of being as 'object-of-knowing' (the only object which we can talk about critically)".[5]

3. The tie between metaphysical realism and epistemological scepticism – I think that the strongest feature of this line of thought descending from Kant lies in the fact that it has brought to light the indis-

[1] See for example Preti 1976b, in particular pp. 143ff.
[2] Preti 1976d.
[3] Preti 1940: 230.
[4] Preti 1943: 119; see also pp 5f, 10.
[5] Preti 1974: 4.

soluble tie between metaphysical realism and epistemological scepticism. As Preti has repeatedly stressed, there can be no overcoming of the sceptical stance as long as we remain committed to a formulation of the problem of knowledge tied to the 'dogmatic' acceptance of a knowing subject opposed to the existence *an sich* "of a real world independent of knowing". It is this ontologically prejudiced approach which legitimises the sceptical stance and makes the problem itself into a contradiction or a 'non-sense'.

Apart from the question of 'non-sense', which I will discuss shortly, for Preti scepticism is nothing but the unfolding of the consequences of the original ontological duplication. In substance, scepticism limits itself to reaffirming "that knowing does not contain its own criterion in itself, that is, it does not contain the criterion of truth and error. *This criterion is external to knowing, it is in something else than knowing*". This is particularly clear if we try to consider the scholastic formula *veritas est adaequatio intellectus et rei* as the *criterion* of truth: "[t]hought does not contain any criterion to establish the *adaequatio*, because it has its own criterion outside *itself*, in the *res*. Reality must be 'seized', 'caught'", but we can never know "whether we have caught a real thing or a shadow". This is why metaphysical realism and the sceptical stance "are simply two sides of the same coin".

We can escape scepticism only if we interpret the sceptical *epoché* as primarily referred to all conceptions which make the criterion of knowing or of truth consist in a transcendent object in itself which is in principle something different from the cognitive representation itself: "[i]f scepticism is realism's self-criticism, it is also the liberation from the realist presupposition". In order to overcome the sceptical stance, epistemology must cast away all metaphysical presuppositions, recovering an "'internal' sense of 'truth' " as "agreement of every *determinate* factual representation and / or *determinate* discursive process to internal criteria of knowledge, such as, precisely, consistency, coherence, etc.".[1]

2.2. KANTIANISM AND THE POSSIBILITY OF METAPHYSICAL REALISM

In the next chapter, I will try to show that the tie between scepticism and metaphysical realism (connected with the elaboration of the Critical notion of truth) has not been affected by recent philosophical developments. In Chapter VI, after having set aside the verifiability principle, I will use this connection, together with moderate epistemic relativism (Chapter II) and the empirical value of our hypotheses and theories (Chapter III), as one of the starting points of my proposal.

Because of the essential role that the connection between scepticism and metaphysical realism will play in my overall argument, I would like first of all to assess its strength, warning against the tendency of

[1] Preti 1974 3,7,12

some supporters of the Critical stance to consider it as a proof of the *senselessness* of metaphysical realism and of the complete *emptiness* of this point of view.

For example, Robert P. Wolff, in his important commentary on the *Critique of Pure Reason*, published in the sixties, stated that Kant's redefinition of objectivity will seem to many as a way not so much of solving as of evading the Cartesian problem of the agreement between the sequences of our subjective ideas and the world of things. According to them, "[u]niversality and necessity may indeed be among the marks of knowledge, but there is also the belief in an object 'out there' (*ob-ject*), standing over against (*gegen-stand*) the subject. If that has been lost, then the result is scepticism, no matter what one calls it".

Of course, "[t]his criticism" expresses the point of view of the metaphysical realist, and Wolff is willing to concede that it is "extremely hard" to withstand. Nevertheless, he considers it "totally without force", because "[t]o some extent it is based on a misunderstanding of Kant's position, a misunderstanding much like that which Dr. Johnson manifested" when he thought he was showing the reality of a stone by kicking it. According to Wolff, "Kant's own statements to the contrary notwithstanding, it is not the teaching of the *Critique* that phenomena are merely in the head, or that in any ordinary sense material objects are not real. But to the more sophisticated objection – that the 'ordinary sense' includes and must include the idea of an ontologically independent object – there can be no answer beyond a careful re-iteration of all the reasons why such a demand is self-contradictory. Universality and necessity are all you can get, Kant says in effect. Therefore, they are all it has ever been legitimate to demand. Anyone who persists in asking what the world is *really* like – in other words, who wishes to know what an object is like independently of the conditions of his knowing what it is like – must then simply be dismissed as unserious. In the terminology of a later philosophical school, he needs to be cured, not answered".[1]

The claim – which is also Preti's – that the Critical-transcendental turn shows that metaphysical realism is a non-sense or a linguistic cramp in need of a cure, seems to me definitely excessive. It is Kant's very perspective which preserves a *meaning* to the opposition between reality in itself and phenomenal reality. Therefore, there is no reason to dismiss as 'unserious', 'in need of cure' or 'not making sense' anyone who believes that our cognitive claims are true only when they agree with the properties of things in themselves, whether or not the criterion with which to establish the agreement is in principle available to us. The realist metaphysician's claim that Kantian epistemology is a form of idealism, incapable of wholly freeing the known object from its dependence upon the knowing subject, is exactly as 'meaningful'.

The proof that we cannot in principle possess a truth criterion which will satisfy the requirements of metaphysical realism and that,

---

[1] Wolff 1963: 322f.

therefore, epistemological correspondentism entails scepticism, cannot be considered a proof of the meaninglessness or of the linguistically pathological character of those requirements. As a matter of fact, Kant's argument is not combined with a conception of meaning which excludes certain assertions as non-significant because they are unverifiable or based upon an improper use of language. The *Restriktionslehre* itself entails the cognitive emptiness, rather than the meaninglessness (or 'blindness'), of categories in their transempirical application.[1] For example, it affirms the illegitimacy of inferring the existence of a reality in itself as the unknown cause of the phenomenon. But obviously it does not show, nor does it claim to show, the meaninglessness of the notion of a reality transcending the known object.

Consequently, unlike Preti, I do not believe that the neopositivist and verificationist approach to the problem of knowledge can be entirely reconciled with that of logical idealism.

As a matter of fact, from a Kantian perspective, metaphysical realism is so far from 'meaningless' that it is possible to formulate the realist objection of the so-called 'neglected alternative' precisely by using the conceptual apparatus of formal idealism, and in particular the distinction between empirical objects and things in themselves.

In the literature on Kant many interpretations have been suggested of the distinction between empirical or phenomenal objects and things in themselves, or noumena. Ultimately, they can be traced back to two fundamental alternatives: the ontological and the methodological interpretation. According to the former, phenomenal and noumenal entities constitute two domains of ontologically distinct entities. According to the latter, there is only one domain of objects, and the distinction concerns two different ways of considering them, namely, in themselves, or under certain epistemic conditions of knowability.

As a matter of exegesis, both interpretations have strong points and weak points. However, it is certain that in both cases it is possible to formulate in a completely meaningful way the realist objection of the neglected alternative. (This objection dates back to Trendelenburg and was later taken up by Hartmann.) The objection contends that Kant's argument for transcendental idealism is vitiated by the unproven presupposition that certain forms of knowledge (space, time, causality, etc.) could only be either structures of the knowing subject, or structures of reality in itself. The neglected alternative is that they could be both these things at once; after all, if things in themselves are unknowable, it is odd to claim that one can know that they do not have the characters of spatiality, temporality, substantiality, causal concatenation, etc.

I agree with Pippin that this objection resists even the refutation recently attempted by Allison.[2] As I will point out shortly, it gains further strength if we consider the possibility of an historical evolution of the forms of knowledge under the pressure of experience. However, it is

[1] Kant 1985: 92f [A51f = B75f]; 173ff [B165ff].
[2] See Allison 1983: 111f and Pippin's review of Allison 1983 (Pippin 1986).

clear that the very fact that this objection can be formulated proves the meaningfulness of metaphysical realism within the Kantian perspective. What this objection makes us understand is that the proof of the possibility of speaking of the object only under certain epistemic conditions of knowability is not tantamount to a proof of the meaninglessness of considering such conditions also as properties inherent in the object in itself, or considered in itself.

In conclusion, from the point of view of transcendental idealism, metaphysical realism (for which space, time and causality are properties of things in themselves, as well as subjective conditions of knowability constitutive of empirical or phenomenal reality) cannot be considered less meaningful than formal idealism (for which space, time and causality are only subjective conditions). To show that such forms are presuppositions of empirical knowledge is not tantamount to showing that they are not properties of reality in itself, and, less still, to showing that statements of that kind are meaningless – exactly in the same way as the logical or empirical non-justifiability of causality and induction, discovered by Hume, considered on its own, cannot be adduced as a proof of the non-existence of causal links and of the non-uniformity of nature, or of the meaninglessness of these statements.

3. THE MATTER OF KNOWLEDGE. EMPIRICIST AND REALIST OBJECTIONS TO TRANSCENDENTALISM

3.1. THE TENSION BETWEEN CORRESPONDENTISM AND COHERENTISM, AND THE DILEMMA OF TRANSCENDENTALISM

I think that the Critical point of view does not show the meaninglessness of metaphysical realism. I suggest furthermore that it does not even establish its *emptiness*. This aspect of the question is interwoven with the problem of empiricism. For the reasons which I have marshalled in favour of moderate epistemic relativism in Chapter II, I hold that radically empiricist and sensationalist theories of knowledge are inadequate. However, I do not think that this can be used to uphold the Kantian theory of synthetic a priori judgements and, consequently, to undermine the idea of the empirically contingent, uncertain and corrigible character of all our synthetic assertions.

I believe that this cluster of problems can be traced back to the internal tensions which have characterised transcendental idealism from the beginning, and which have been fully brought to light by post-Kantian scientific and philosophical developments. It is possible to identify these tensions starting with a reflection on the tendency of the Critical theory of truth to maintain itself in an intermediate position between correspondentism and coherentism.

The Critical theory of truth is not a pure coherence theory, since it accepts the nominal definition of truth as agreement between knowledge and its object, and it makes the criterion for the discrimination of true from false depend not only upon formal components, but also upon given material components.

On the other hand, the Critical theory of truth is not a correspondence theory either, since, in subordinating the evaluation of judgements to 'rational' requirements, it makes knowledge and truth something different from the purely passive 'copy' of a given object, wholly independent of the subjective conditions of knowability.

As I have said, it is the insistence on both components of the cognitive process which makes the Critical theory of truth particularly suitable to the network model described in Chapter III.[1] The network model, as does the Critical theory, bestows equal importance upon rational systematicity and empirical adequacy.

At the same time, however, the mixture of correspondentism and coherentism makes it difficult to consider as fully warranted transcendental epistemology's anti-empiricist and anti-realist claims.

In order to clarify the problem, I will look at the dichotomy between form and matter of knowledge from the point of view which Herder felicitously called 'meta-critical'.[2] At this extremely general level, *matter* is the determinable in general and that which in appearance corresponds to sensation; *form* is the determination of the determinable and that which organises sensible matter in a certain way producing the phenomenal object.[3]

But, whereas the formal components of knowledge can receive some form of specification because they basically coincide with the a priori judgements (analytic and synthetic) and, possibly, with the regulative principles of reason, nothing as specific can be said of the material components. Certainly, they cannot be identified with synthetic a posteriori judgements, since synthetic a posteriori judgements, in so far as they are expressions of empirical knowledge, are syntheses of form and matter. Nor can they be specified by reference to the "judgements of perception", which in paragraphs 18-19 of the *Prolegomena* are opposed to the "judgements of experience", for, as it has repeatedly been shown, even judgements of perception contain a formal element. Some interpreters have declared the distinction itself to be downright incomprehensible from a transcendental point of view.[4]

Actually, the question of the matter of knowledge makes one strongly feel the singular destiny of a philosopher who, on the one hand, criticises Locke for having "*sensualised* [...] concepts" and Leibniz for

---

[1] It is possible to classify Kant's conception as a 'theory of truth as conformity to a rule or norm of judging'; Walker makes use of the expression "impure coherence theory" (Walker 1989: 63).

[2] See Beck 1978: 20ff, as well as Pippin 1982: ix. The critical remarks that follow are developed after Parrini 1994a.

[3] See Kant 1985. 280 [A266 = B322]; 65f [A20 = B34]; and sec. 15, 'The Possibility of Combination in General', of the second version of the 'Transcendental Deduction of the Pure Concepts of the Understanding' (Kant 1985· 151f [B129f])

[4] See for example Cassirer 1981: 158 n. 20, Allison 1983: 151f; Guyer 1987: 99f.

having "*intellectualised* appearances",[1] but, on the other hand, is forced to introduce schemes to mediate between sensibility and pure concepts of understanding, without being able to specify the pure sensible material contents of knowledge.

For this reason, Kemp Smith in the end considers the author of the *Critique* the "founder" of the coherence theory of truth, despite his recognising that Kant "never himself employs the term Coherence" and "constantly adopts positions which are more in harmony with a *Correspondence* view of the nature and conditions of knowledge". According to Kemp Smith, the decisive point is that in transcendental epistemology "the test of an asserted fact, *i.e.* of our description of a given appearance, is its conformity to the principles that make insight possible": "[n]ot contents alone, but contents interpreted in terms of some specific setting, are the sole possible objects of human thought".[2] As Cassirer expressed it, "all the determinateness, that we can ascribe to the 'matter' of knowledge, belongs to it only relatively to some *possible order* and thus to a formal *serial concept*. The particular, qualitatively specific sensation gains its character only by distinction from other contents of consciousness; it exists only as a serial member and only as such can be truly conceived. Abstraction from this condition would result not merely in a greater or less 'indefiniteness' of its content, but would absolutely annihilate it".[3]

Despite this, what can be said in general about the dichotomy between form and matter of knowledge is sufficient to raise well-grounded realist and empiricist perplexities about transcendental idealism. These perplexities have a history of their own in Kantian literature and can be ultimately traced back to the question of 'determined' empirical knowledge. Kant poses the question of the possibility of synthetic a priori knowledge, and through it of knowledge in general; but it is clear that an adequate solution of the latter problem cannot be separated from an explanation of the possibility of determined empirical knowledge. Instead – as Broad pointed out – one of the most surprising aspects of the *Critique* and of the *Prolegomena* is that there is never any question about "what it is that determines the particular shape, size, and position which a particular object is perceived to have on a particular occasion".[4] The problem has been raised also by other authors (for example, C.I. Lewis), especially in order to show that subjectivist idealism is not capable of accounting for the particularity of experience.[5] It is not a coincidence that an incisive formulation of the question can be found

---

1 Kant 1985: 283 [A271 = B327].
2 Kemp Smith 1979: xxxviif, 36.
3 Cassirer 1953a: 310f.
4 Broad 1978: 25
5 Lewis 1956: 189ff.

already in Herbart's philosophy, where it is connected with the distinction between phenomena and things in themselves.[1]

The problem issues from the fact that an essential feature of Kant's conception, well reflected by the Critical theory of truth, is the fact that the cognisant subject can be considered the 'legislator of nature' only as regards the formal aspects of nature itself (*natura formaliter spectata*). Empirical concepts, natural laws and synthetic a posteriori judgements in general (*natura materialiter spectata*), even while having to be compatible with synthetic a priori principles, are not derivable from them: they depend on experience. In science, there is an a priori part that constitutes the (Critical) metaphysics of every science of nature, and an a posteriori part that constitutes its true and proper physics.[2] The relations between these two parts are complex, and it is not always easy to say where exactly Kant sets the border between them.

It is certain, however, that natural science presupposes the formal, purely a priori components of knowledge, as much as those which depend on experience. The formation of the concepts and laws of natural science would be impossible without the determinant contribution of experience. For example, given the transcendental validity of the causal principle, if "wax, which was formerly hard, melts, I can know *a priori* that *something*" (for instance, the heat of the sun) "must have preceded, [...] upon which the melting has followed according to a fixed law". But "independently of experience, I could not determine, *in any specific manner*, either the cause from the effect, or the effect from the cause": "[t]hat sunlight should melt wax and yet also harden clay, no understanding [...] can discover from the concepts which we previously possessed of these things, much less infer them according to a law. Only experience is able to teach us such a law".[3]

It seems to me, however, that it is precisely this dependence of 'determined' knowledge on experience which presents the Critical-transcendental conception with a dilemma to which I see no solution.

(1) Let us suppose that the matter of knowledge is something completely ...material, i.e., totally formless and structureless, a sheer determinable. On this basis, one can still hope to explain (excluding other sorts of difficulties) the possibility of synthetic a priori judgements, or of justifying the validity of a set of universal and necessary forms of the knowledge of objects of experience. But then it becomes difficult to understand how a matter of knowledge totally devoid of inner connections and articulations can 'channel' our interpretative and constitutive capacities, allowing us to form concepts and empirically determined scientific laws.

---

[1] Herbart formulated the question in sec. 150 of the fourth edition of Herbart 1964 (sec. 127 in the first edition); see in particular p. 211. See also proposition 4.0412 of Wittgenstein's *Tractatus*, where the famous spectacles metaphor appears (Wittgenstein 1994: 23).

[2] Kant 1985: 172f [B163ff].

[3] Kant 1985: 610 [A766 = B794]. See also 237 [A216 = B263].

(2) If on the other hand we attribute a certain 'form' to matter in order to account for the role of experience in the development of knowledge, the question rises of the knowability of this structure taken by itself. Above all, it becomes difficult to understand, despite Kant's demonstrative efforts in the 'Transcendental Analytic', why experience cannot have some 'determinant' effect on the supposed a priori parts of knowledge as well as on the a posteriori ones.

Let us take, for example, the case of geometry. Poincaré maintained that experience cannot decide the question of the validity of physical geometry. This is because "[i]n space, we know" both "rectilinear triangles the sum of whose angles is equal to two right angles", and "curvilinear triangles the sum of whose angles is less than two right angles". Since to accept Euclidean geometry would mean "[t]o give the name of straights to the sides of the first", and to adopt a non-Euclidean geometry would mean to "give the name of straights to the sides of the latter", it is clear that we cannot accept the thesis of the empirical foundation of the spatial metric.[1]

On first impression, these words sound somewhat strange, since one fails to see how the choice of a geometry can depend on calling a rectilinear triangle curvilinear and a curvilinear triangle rectilinear. However, setting aside Poincaré's far from trivially groundless claims, the example can be adapted to suit our problem. If in experience there are triangular figures (it is not necessary to specify whether curvilinear or rectilinear), the sum of whose internal angles is equal to two right angles, the determination of the Euclidean or non-Euclidean structure of the regions of space to which such figures belong is not in principle different from the determination of the rectilinearity of these triangles (here it is not relevant whether it is possible to carry out this determination by making use of experience alone). And if it should result that the rectilinear triangles are those the sum of whose internal angles is different from two right angles, with the same operation of empirical testing (however structured), determined knowledge would have been attained, concerning at the same time the rectilinearity of some triangles and the non-Euclidean geometrical structure of some portions of space. The point is that the empirical-experimental procedures by which it is possible to attain determined 'synthetic a posteriori' knowledge about specific aspects of 'objects of experience' (the rectilinearity of a triangle), are the same procedures by which it is possible to assess the validity of 'synthetic a priori' principles regarding the most general aspects of 'empirical objectivity' (the geometrical structure of space).

3.2. CRISIS OF NEWTONIANISM AND CRISIS OF KANTIANISM

It is not a coincidence that the crisis of the synthetic a priori has manifested itself in connection with the crisis of a 'determined' theory of

[1] Poincaré 1946c: 235

physics, namely Newton's. This did not happen, as some have improperly suggested, because Kant had sought to establish the universal and necessary validity of that physics as an empirical theory. It happened instead because the crisis of Newtonian physics affected also principles, such as Euclideanism and causality, which in the *Critique* and in the *Prolegomena* were counted among the 'metaphysical' presuppositions (in the sense of 'Critical metaphysics') of any possible science of nature (therefore also of Newton's). The supposed a priori metaphysical components turned out to be 'holistically' interwoven with a posteriori physical ones, and no less vulnerable than the latter with regard to experience.

The dilemma is analogous to the one I described in the previous chapter discussing the relation between observation and theory. It is true, as Cassirer says, that it is not possible to specify material contents without giving them some formal structure; but we cannot because of this maintain that the 'sensible manifold' is totally malleable by thought, without making the 'sensible manifold' automatically useless for the construction of knowledge. On the other hand, we cannot recognise a certain autonomy to perception with respect to reason, without exposing also the rational principles to the danger of refutation by experience. Personally, I do not see how we can attribute a determining function to the sensible material if we fail to acknowledge that observation – as prescribed by the network model – is capable of putting us in front of data of experience which are sufficiently autonomously structured as to introduce some limitations in the range of admissible alternative interpretations.

The point was already clearly made by Reichenbach in 1920, when he observed that "[t]he ultimate instrument in the judgment of empirical truths, *perception*, is *independent* of reason; although perception always permits different *interpretations*, the combination of the interpretations is no longer arbitrary. It is the significance of the theory of relativity to have discovered the limits of arbitrariness".[1] The theory of relativity shows clearly, for example, the empirical impossibility of upholding a system of interpretative principles to which belong simultaneously Euclidean geometry and a definition of congruence based on the so-called 'rigid rod'. The objection can be formulated in general terms by saying – as Reichenbach himself did in 1933 – that "for every totality of experience there exists a system of pertinent preconditions, but, conversely, for every given system of preconditions a totality of experience can be constructed that contradicts this system of preconditions".[2]

In order to salvage Kantianism, Cassirer reinterpreted the synthetic a priori principles as the "*ultimate logical invariants* [...] which lie at the basis of any determination of a connection according to a law". He further maintained that "even if the particular principles named by Kant are not the ultimate ones, the possibility remains that

[1] Reichenbach 1978e: 38.
[2] Reichenbach 1978d: 401.

fundamental preconditions nonetheless do exist and must be gradually figured out".

This 'liberalisation' of the synthetic a priori is empty and unjustifiable. It postulates the existence of principles which it is not able to specify, and it does not avoid the previous objection "that it can never be said of any system of synthetic principles that it will be consistently compatible with all future experience".[1] It is right to maintain that scientific activity takes place within a framework of presuppositions; but we must also recognise, in an empiricist spirit, that the presuppositions adopted from time to time depend holistically upon experience, and that they too are uncertain and corrigible.

Some of Kant's late formulations seem to offer a useful starting point for the specification of the connections between empiricism, the crisis of the transcendental interpretation of the synthetic a priori, and the contextual a priori, which I defended in Chapter II as an integral part of epistemic relativism. Granting a certain interpretation[2] of the very problematic notes on the passage from metaphysics to physics which go under the name of *Opus postumum*, Kant seems to endow the understanding not only with a certain categorial apparatus, but also with the capacity of freely "inventing" [*erdichten*] conceptual constructs not directly given in sensible experience. And, in a remarkable passage of that work, he states that it would be "dangerous" for the correct foundation of science if the assertions of physics were to get mixed up with the metaphysical principles of the science of nature, insinuating themselves into the system of such principles. If this happened, the "*a priori* and empirical principles might communicate with or *make claims upon one another*".[3] These are truly prophetic words: what happened is exactly what Kant feared. The understanding has utilised its free conceptual constructions, such as, for example, the notion of rigid rod at the basis of measurement, in order to set experience against one of the a priori forms of knowledge, Euclidean geometry.

For this reason, in arguing for the distinction between a priori and a posteriori and between analytic and synthetic in Chapter II, I stressed the purely relative and contextual value of these distinctions, denying that they can be founded upon an absolute, logical or naturalistic, basis. We can and must preserve the meaning of these distinctions; but this meaning must strip them of any a-historical absolute character, connecting them with the relatively stable presuppositions on which the evaluation of our judgements depends. Only thus may the a priori (analytic and synthetic) be also freed from any connection with the constitution of empirical subjectivity, since we can no longer take for granted the identification of the physiological, psychological, sociological and anthropological conditions of our experience of objects, with the

[1] Cassirer 1953a: 269.
[2] See for example Mathieu 1958.
[3] Kant 1993: 37 [21: 526], italics added

contextually a priori conditions which constitute the framework for the evaluation of our particular cognitive claims.

3.3. MATTER OF KNOWLEDGE, THINGS IN THEMSELVES, AND METAPHYSICAL REALISM

On the other hand, the historical mutability of the network of epistemic conditions, their global empirical responsibility, and the purely contextual value of the distinction between a priori and a posteriori, are all relevant to the relations between logical or formal idealism and metaphysical realism. Herbart, raising the problem of determinate knowledge, had already observed that there must be some relation between things in themselves, the matter of knowledge and the properties which we attribute to phenomenal objects on the basis of experience.[1]

In this century, the discussion of this aspect of Kantianism has intertwined itself with the vicissitudes of analytic philosophy. The theoretically relevant aspect of the matter was well formulated by Schlick, but it was noted also by Russell[2] and by Wittgenstein:[3] "if the 'phenomena' are appearances of something else, then the mere fact that this 'something else' is that particular reality of which that particular phenomenon is the appearance – this fact enables us to describe the reality just as completely as the appearance of it. The description of the appearance *is*, at the same time, a description of that which appears. The phenomenon can be called an appearance of some reality only in so far as there is some correspondence between them, they must have the same multiplicity; to every diversity in the phenomenon there must be a corresponding diversity in the appearing things, otherwise the particular diversity would not form part of the phenomenon qua phenomenon, nothing would 'appear' in it. But if this is so, it means that the 'appearance' and the 'appearing reality' have identically the same structure. The two could be different only in content, and as content cannot possibly occur in any description, we conclude that *everything* which can be asserted of the one, must be true for the other also. The distinction between appearance and reality collapses, there is no sense in it".[4]

Of course, a defender of Kant, faced with Schlick's attempt to abolish the distinction between phenomenal appearance and reality-in-itself, could point out that Schlick's attempt depends on an outmoded form of the principle of verifiability, and that one of the principal teachings of the *Critique* remains the demonstration of the inevitable conditioning exercised on the known object by certain forms of knowl-

---

[1] See Herbart 1964: 211.
[2] See Russell 1950: 61f.
[3] See proposition 4.0412 of the *Tractatus*, where the spectacles metaphor appears (Wittgenstein 1994: 23). See also Broad 1978. 25.
[4] Schlick 1979g: 359.

edge, whereby it will never be possible to know what things in themselves are like. But the huge literature on transcendentalism is there to prove that, even if we place ourselves in a strictly Kantian perspective, there still remains the problem of somehow explaining the relation between phenomenal appearances and things in themselves. And it is difficult to consider realism's requests completely empty or unjustified, if we do not possess a theory of meaning powerful enough to warrant the meaninglessness of realism – all the more when we are forced to recognise the changeability of the presuppositional system.

It will always be possible to contest the legitimacy of projecting the formal and material components onto an ontological background as two entities existing in a separate and opposed manner, by recurring to Cassirer's theory that "[m]atter *is* only with reference to form", and that form "*is valid* only in relation to matter". Analogously, we could follow Cassirer again, in disqualifying, as unacceptable applications of the categories to objects transcending the field of possible experience, the old questions about the status of cognitive conditions: if they are "an element of *being*", or of nature, or "mere constructions of *thought*", or "the universal forms of expression of our consciousness".[1] But, as I have said, such objections cannot be equated to a proof of the senselessness of metaphysical realism, and not even of its compete emptiness.

If we acknowledge that cognitive presuppositions can be modified under the pressure of conceptual innovation and of the data of experience, we end up with unpleasant consequences for Kant's original aspirations. On the one hand, it will no longer be possible to warrant cognitive objectivity by appealing to the universal and necessary character of the principles of knowledge. On the other hand, the tendency to interpret scientific change as the progressive approximation to the knowledge of a transcendent reality will gain momentum.

Recent events, marked by variously motivated significant reformulations of metaphysical realism, and by the diffusion of anarchist and conversational renunciations of cognitive objectivity, exemplify very well both horns of the dilemma. I would like to outline an alternative to these two extreme positions by means of positive philosophy. In order to do this, I will have to consider beforehand if, and to what extent, the most recent developments in the philosophy of science have confirmed the connection between scepticism and realist metaphysics. Only after a positive solution of this question will I be able to show how it is possible to preserve a certain dose of objectivity in our knowledge, while rejecting every form of foundationalism, be it empiricist, rationalist or realist.

---

[1] Cassirer 1953a: 311, 309.

Chapter V

REALISM, INSTRUMENTALISM, NEOPOSITIVISM

1. REALISM VS SCEPTICISM

In this chapter, I will discuss the problem of metaphysical realism, of the existence of a reality in itself, an absolute reality constituting the transcendent counterpart of our cognitive acts. I will connect the problem of metaphysical realism to the opposition between scientific realism and instrumentalism specific to the philosophy of science. By 'scientific realism' and 'instrumentalism', I refer to two different positions on the problem of the existence of the 'theoretical' entities posited by science. The scientific realist maintains that there is no reason to assume a different ontic attitude to the existence of objects such as chairs and tables on the one hand and electrons and neutrinos on the other: from a general point of view, apart from specific doubts, both kinds of entities have an equal claim to existence. On the contrary, the instrumentalist contends that only observable objects really exist, while theoretical entities are nothing else but useful calculating devices – 'representative fictions', as Bain said[1] – introduced in order to organise and to predict experience.

Metaphysical realism is usually discussed in works on the theory of knowledge, while scientific realism is usually discussed in works on the philosophy of science. The reason why I keep the two questions together is that I intend to argue for the reality of both theoretical and observable entities, but in both cases in a non-metaphysical sense of the term 'reality'. In brief, my thesis is that the instrumentalist is wrong in thinking that theoretical entities are epistemologically problematic, while observable objects are not. My conclusion will be that, apart from specific doubts (in the sense that will be clarified in the final chapter) both observable and non-observable entities must be considered real; but their reality is to be understood in an empirical or scientific sense, and not in a metaphysical sense.

As Richard Boyd, a prominent champion of realism, has pointed out, there is "a single, simple, and very powerful epistemological argument" on the basis of which anti-realist philosophers have constantly rejected realism. Boyd states it with respect to scientific realism but, in my view, it really refers to metaphysical realism, and it concerns scientific realism only in so far as scientific realism is intended in a metaphysical sense. The argument can be summarised as follows. Given any theory which talks about entities, attributes and relations which go beyond what is immediately given in experience, from a strictly logical point of view a plurality of alternative theories is always possible, which

[1] Bain 1870, vol. II· 132, see p. 364 for the case of 'representative fictions' in medicine.

fit with the same available empirical data, while making different statements about non-immediately given reality. It follows that experience can never guarantee the correctness of certain statements rather than others. Many sceptical arguments (for instance, those based on perceptual illusions) are a variant of this line of reasoning.

The standard instrumentalist objection against scientific realism simply adapts the argument as follows. Granting the existence of observable objects, the reality of the unobservable entities posited by a scientific theory *T* in order to explain the properties of the observable entities is questioned, since it is always logically possible to construct a plurality of alternative theories which, though they are "empirically equivalent" to *T*, "offer contradictory accounts of the nature of unobservable phenomena. Since [...] *T* and each of the theories empirically equivalent to it will be equally well confirmed or disconfirmed by any possible observational evidence", "no scientific evidence can bear on the question of which of these theories provides the correct account of unobservable phenomena; at best, it might be possible to confirm or disconfirm the claim that each of these theories is a reliable instrument for the prediction of observable phenomena. Since this construction is possible for any theory *T*, it follows that scientific evidence can never decide the question between theories of unobservable phenomena and, therefore, knowledge of unobservable phenomena is impossible".[1]

1.1. DUHEM'S 'PASCALIAN' STANCE

This kind of argument is closely connected with the aristocratic problem of induction and with Duhem's conception of the logical structure of empirical testing, which I discussed in Chapter II, sections 2.2. and 2.3., Duhem was always well aware of the anti-realist implications of his critique of inductivism and of the theory of crucial experiments. However, at the same time he constantly emphasised that the historical evolution of science, in particular of physics and astronomy, reinforces the scientists' conception of the development of science as the progressive uncovering of the ultimate characteristics of reality. Duhem thus presented 'the aim' of theories in an intentionally ambivalent way.

(1) On the one hand, he denies the possibility of demonstrating that theoretical constructions grasp the reality underlying sensible appearances and explain the experimentally testable phenomenal regularities. "To explain (explicate, *explicare*) is to strip reality of the appearances covering it like a veil, in order to see the bare reality itself".[2] This task 'transcends' the methods of observation utilised by physics, and can be pursued only by metaphysics. The methods of physics restrict science to the domain of empirically observable phenomena and their nomological connections. Physicists could lay claims to the explanation

---

[1] Boyd 1984: 42, 44.
[2] Duhem 1962a: 7.

of phenomenal regularities only by advancing into the metaphysical field.[1]

If they want to avoid metaphysics, they will have to look at their theories not as explanations, but as systems "of mathematical propositions, deduced from a small number of principles [fundamental hypotheses] which aim to *represent* as simply, as completely and as exactly as possible a [large] set of experimental laws".[2] For Duhem, as for Mach, the fundamental task of physical theory is to contribute to a form of "intellectual economy", by means of substituting a narrow number of principles for a very large number of unrelated laws, each one of which, were it not for the theory, would have to be learnt and considered on its own.[3]

(2) On the other hand, Duhem's methodological conception induces him to argue that all theoretical components, even the more abstract ones, can undergo the scrutiny of experience, being integral parts of a theoretical system which, as a whole, is empirically testable. Although empirical observation does not impose a univocally determined correlation between certain experiences and certain statements of the system, nevertheless, empirical observation plays a fundamental role in the overthrowing of theoretical hypotheses and their substitution by empirically more satisfactory theories. Moreover, the historical evolution of scientific knowledge would seem to buttress our idea of the descriptive and ontological value of physical theories.

Duhem thus assigns to theoretical elaboration a *classificatory* function endowed with an ontological import exceeding the purely *economical* and *representational* function: "[w]ithout claiming to explain the reality hiding under the phenomena whose laws we group, we feel that the groupings established by our theory correspond to real affinities among the things themselves". The more the physical theory perfects itself, the more we *feel* that the *logical order* in which it organises the experimental laws *reflects an ontological arrangement.* Of course, the physicist is not able to demonstrate that "the logical order in which theory orders experimental laws is the reflection of an ontological order [...] transcending experience", since "the method at his disposal is limited to the data of observation".[4] Nevertheless, the physicist cannot escape the conviction that a theoretical system resembling so closely "a work of art" for its capacity of introducing a *simple, harmonious* and *beautiful* order in an enormous quantity of initially very disparate laws, is not "a purely artificial system", but a proper "*natural classification*".[5]

---

[1] See Duhem 1962a: 10.

[2] Duhem 1962a: 19.

[3] Duhem 1962a. 21, 21, n. 1 (with reference to Mach, 'On the Economical Nature of Physical Inquiry' [Mach 1986]).

[4] Duhem 1962a· 26f.

[5] Duhem 1962a: 24, italics added.

Duhem, then, recognises that the epistemological analysis of physical theories leads to a dilemma: he believes that we can do nothing better than appeal to Pascal's "reasons of the heart 'that reason does not know'". "[T]he analysis of the methods by which physical theories are constructed proves to us with complete evidence that these theories cannot be offered as explanations of experimental laws; and, on the other hand, an act of faith, as incapable of being justified by this analysis as of being frustrated by it, assures us that these theories are not a purely artificial system, but a natural classification. And so, we may here apply that profound thought of Pascal: 'We have an impotence to prove, which cannot be conquered by any dogmatism; we have an idea of truth which cannot be conquered by any Pyrrhonian skepticism'".[1]

2. THE CRISIS OF POPPER'S FALLIBILIST REALISM

A. Popper's deductivist attempt to overcome scepticism

A few decades later, Popper went in search of a much more ambitious solution to the problem of scepticism than Duhem's. If Popper's arguments were sound, we would have to recognise that we are much less incapable of demonstrating facts about the adequacy of theories than Pascal and Duhem had supposed. In order to clarify the point that is of interest here, it will be sufficient to recall the connections between the theory of corroboration, the method 'of bold conjectures and severe attempts at refutation', and the 'theory of truth and of verisimilitude'. In Popper's view, it is these connections that enable reasons 'not of the heart' to sustain realism and overcome Hume's irrationalism and scepticism.

Popper believes that it is impossible to definitively establish the truth or the falsity of our statements, even in the case of the test statements. Accordingly, if the term 'knowledge' is intended in the sense of definitive, absolutely certain knowledge, then we should say that in the empirical sciences there can be no knowledge whatsoever: "*[w]e do not know: we can only guess*".[2] The historically and theoretically important point is that, in Popper's fallibilism, the acceptance of the conjectural character of all our assertions is not supposed to entail a form of masked scepticism which assigns the same epistemic value to all our cognitive claims. Popper thinks he has been able to devise purely logical-deductive criteria, uncompromised by the abhorred inductive method, with which to evaluate the comparative merits of our hypotheses and theories. To be more precise, those rules should enable us to choose, from the infinite set of logically possible assertions, those which 'maximise' the risk of "clashing with reality".[3] Those of such asser-

---

[1] Duhem 1962a: 27.
[2] Popper 1977a: 278.
[3] Popper 1974a: 981.

tions which survive empirical tests can be reasonably conjectured to be closer to the truth than the others.

The idea that supports this proposal – inadequately appreciated in its day even by a philosopher of science very close to Popper such as Lakatos[1] – hinges on the thesis that *risky* predictions of a *false* theory have a higher logical probability of failing than *non*-risky ones. Thus, we would have a deductively defensible reason to conjecture that *severe* tests have more logical probability of eliminating false theories than non-severe tests.

According to Popper, the mathematical theory of probability guarantees a priori that the predictions of a theoretical hypothesis are more daring and risky the larger the assertive content of the hypothesis itself. It follows that in science we should systematically search for the theory comparatively richest in content, in order to subject it to the most severe attempts at refutation of which we are capable: "[w]e try to find its weak points, to refute it. If we fail to refute it, or if the refutations we find are at the same time also refutations of the weaker theory which was its predecessor [...], then we have reason to suspect, or to conjecture, that the stronger theory has no greater falsity content than its weaker predecessor, and, therefore, that it has the greater degree of verisimilitude".

The Popperian method bases itself on:

(1) the theorem that, with the increase of the logical content of a theory, its truth content also increases;

(2) a definition of the concept of verisimilitude, according to which, "[i]n brief", a theory "T_2 is nearer to the truth, or more similar to the truth, than T_1, if and only if more true statements follow from it, but not more false statements, or at least equally many true statements but fewer false statements".[2]

The work carried out by Popper in order to specify in a logically rigorous way the qualitative and quantitative aspects of verisimilitude could also be used to dispel Quine's perplexities on the practicability of Peirce's idea of defining "truth outright in terms of scientific method, as the ideal theory which is approached as a limit when the (supposed) canons of scientific method are used unceasingly on continuing experience".[3]

Popper's conception is meant to give us some good purely deductive reasons to establish when a conjecture has better chances than another of being closer to the truth. Supposedly, these reasons are purely deductive, because all the notions employed, from falsification to riskiness, from logical content to probability, can be established a priori by means of logico-analytic reasoning which bars all reliance upon the kinds of verificationism and inductivism that lead directly to Humean scepticism.

[1] See Popper 1974d: 1101-3; 1974b: 1010-3; Parrini 1980c: 143-5 n. 57.

[2] Popper 1975b: 52f (the original formulation is in Popper 1972b)

[3] Quine 1970. 23.

As for the notion of truth, this is defined according to the model suggested by Tarski. However, contrary to Tarski,[1] Popper explicitly attributes an epistemological value both to the notion of truth and to the derived notion of verisimilitude. Both are assumed to be logically rigorous codifications of the philosophical idea at the basis of naïve and sophisticated realism, namely, "that truth is correspondence with the facts (or with reality); or, more precisely, that a theory is true if and only if it corresponds to the facts".[2] And a theory is more verisimilar than another when it comes closer than the other to grasping the traits of reality.

Both notions, taken in a correspondentist, objective and absolute sense, must be assumed as regulative ideals in the Kantian sense, guiding scientific research and critical-rational discussion.[3] Although there are no *criteria* to establish with certainty when truth has been attained or when we have come closest to it, by using these notions and the deductivist and falsificationist methodology of bold conjectures and rigorous attempts at refutation it will be possible to attain good reasons to *conjecture* that certain hypotheses or theories (for example: relativistic physics), have more chances than others (for example: Newtonian physics) to attain the truth or to come closer to it. Thus Popper's fallibilism supposedly overcomes the Pyrrhonian aspect of Duhem's thought, lending rational support, based upon solid deductive arguments, to Pascal's "reasons of the heart".[4]

B. Objections to Popper

Unfortunately, as was the case with neopositivism, Popper's conception has not survived rigorous analytic scrutiny carried out on various levels. It collapses under the blows of weighty logical, methodological and factual objections, which dispute its descriptive adequacy as much as its normative value. All these objections seem to lead to the same general conclusion, namely the impossibility of overcoming scepticism by adhering to Popper's peculiar metaphysical realism. Moreover, these objections do not seem to me to be affected by Popper's suggestion to kill, as it were, justificationism by hurling at it the insult of meaninglessness.[5] Since these objections are very well known, I will only recall their gist.[6]

1. The problem of the definition of verisimilitude – Certain logical objections affect the concept of verisimilitude abstractly considered: it seems that Popper has not been able to define in a logically correct way a notion of approximation to truth which allows us to hierarchise false theories according to their greater or lesser content of truth and / or

---

[1] Tarski 1964: 33f.

[2] Popper 1975b: 44.

[3] Popper 1972b, secs VIII and IX; 1974a. 15, 226, 1157; 1975a: 29f, 120, 317f; 1983: p I, ch. I, sec. 2 / III

[4] See Popper 1975b: 58f; 1972b: 224-28; 1974d: 1101-3; 1974b: 1010-3.

[5] See Popper 1983. 21.

[6] For a more extensive discussion, with bibliographical references, see Parrini 1980c.

falsehood. Indeed, it is controversial and still a subject of debate whether it is possible to codify such a notion in a formally rigorous way.

In the 1982 Introduction to the *Postscript to the Logic of Scientific Discovery*, Popper acknowledges the logical defects of his notion of approximation to truth. He counters that "a formal definition of verisimilitude is not needed for talking sensibly about it".[1] If this were the case, a supporter of inductive confirmation should also be allowed to set aside the Humean problem of the justification of induction, and be satisfied with the common intuitive belief in experience's capacity of confirming our empirical generalisations. Yet Popper will not allow this. This partiality seems to me quite incompatible with the critical-rational discussion Popper advocates.

However, this debate concerns only the logical aspect of Popper's position. Even if the question about the possibility of a formally rigorous codification of verisimilitude were answered in the positive, two further problems, both essential to the defence of metaphysical realism against epistemological scepticism, would remain open. These are the problem of the intuitive plausibility of the concept thus obtained, and the problem of the epistemic criteria necessary for its use in the evaluation and the selection of scientific theories.

2. *Methodological and epistemological aspects of verisimilitude* – Other objections concern the logical, methodological and epistemological premises of the arguments for the application of the notion of verisimilitude and for the defence of the method of bold conjectures and severe attempts at refutation. A number of authors have contributed to the formulation of these criticisms, including Grover Maxwell, Wesley Salmon and Adolf Grünbaum.

Grünbaum, in a famous series of papers published in the seventies, exposed three major defects of Popperian methodology:

(a) Given the existence of a considerable logical incompatibility between the critique of inductivism and the quantitative theory of content and verisimilitude, both inductivism and the quantitative theory of content and verisimilitude turn out to be equally problematic from the point of view of the probability calculus.

(b) The judgements of verisimilitude depend on the tacit use of the *bête noir* of Popper's entire epistemology, namely, induction by enumeration. Roughly: in order to infer the closer approximation to truth of one theory over another on the basis of its performing better in empirical tests, we are compelled to use past results as a basis for an inductive extrapolation concerning the outcomes of future tests.

(c) The argument according to which "the *risky* predictions of a *false* theory are more likely to fail than its *non*-risky ones" (intended to supply rational-deductive "grounds for conjecturing that severe tests have a better chance of weeding out false theories than non-severe tests") is not valid.

[1] Popper 1983: xxxvi.

Objections (a) and (b) can be considered in a sense 'retaliatory': they show that Popper's deductivism is no better founded than the inductivism he disparages, since Popper's deductivism can function only if we acknowledge the legitimacy of inductive methodology by accepting its basic principle of the *straight rule*.

On the other hand, objection (c) affects the very heart of fallibilism. It has particular significance from the point of view of the epistemic relativism defended in Chapter II. Objection (c) starts from the fact that, in Popper's formulation, a theory's predictions are more or less risky with regard to the hypotheses which constitute the 'background knowledge' in a particular historical moment of the evolution of science. Consequently, this objection can be used to show that there is no absolute criterion with which to judge if a theory is more risky than another, and if a test is more severe than another. Evaluations of riskiness and severity are relative to an historically determined scientific context, and they change with the changing of our knowledge. The comparative degrees of riskiness and severity depend upon accidental historical circumstances; they are not absolute logical properties. Therefore, either Popper's evaluation criteria do not allow us to say anything about a theory's greater chances of truth or falsity, or they allow us to say something about the matter only relatively to the epistemic conditions by means of which we judge the riskiness and severity of tests and theories.

From this point of view, it is possible to point out a further 'retaliation', this time not affecting induction by enumeration, but eliminative induction: Popper's deductivism, as a means of maximising our chances of eliminating false theories, thereby leading us to the remaining true theories, is no better than Bacon's, Whewell's and Mill's eliminative inductivism.[1]

3. The descriptive value of verisimilitude – Finally, there are objections which concern the applicability of the notion of approximation to truth. Unlike the previous ones, they concern the descriptive, rather than the normative, claims of fallibilism: the target of these objections is not the alleged conditions of a deductively sound use of the concept of verisimilitude, but the possibility of using verisimilitude to describe the factual characteristics of science and scientific change.

The 'factual' objections highlight the problematic character of an *intertheoretical* or *intersystemic* notion of truth, applicable to sentences independently of a conceptual scheme (the problem was already addressed by Quine in *Word and Object*). These objections dispute the possibility, or even the '*meaningfulness*', of a description of scientific development formulated in terms of approximation to an hypothetical true theory corresponding to reality.

---

[1] Grünbaum 1976: 106, 108. For Salmon's position on the relationships between Popper's conception of corroboration and "the problem of rational prediction in contexts of practical decision making", see Salmon 1988.

According to Quine, all sentences, except those which are "pretty firmly and directly conditioned to sensory stimulation" are "meaningless intertheoretically". That is, they have a meaning only "relative to" their "own theory". Therefore, even the application of the notion of truth cannot but be relative to a background theory. It is only "when we turn back into the midst of an actually present theory, at least hypothetically accepted, that we can and do speak sensibly of this and that sentence as true". It is 'meaningless' to speak in an absolute way of the truth value of an individual sentence: "[w]here it makes sense to apply 'true' is to a sentence couched in the terms of a given theory and seen from within the theory, complete with its posited reality. Here there is no occasion to invoke even so much as the imaginary codification of scientific method. To say that the statement 'Brutus killed Caesar' is true, or that 'The atomic weight of sodium is 23' is true, is in effect simply to say that Brutus killed Caesar, or that the atomic weight of sodium is 23".[1]

This kind of objection becomes even stronger when, with the new philosophers of science, we assert the strong discontinuity of the history of science, and we introduce the notion of incommensurability. Kuhn criticised those philosophers who, though "[g]ranting that neither theory of a historical pair is true, [...] nonetheless seek a sense in which the later is a better approximation to the truth", and "wish [...] to compare theories as representations of nature, as statements about 'what is really out there'". The history of science does not support this idea in the least. "To say, for example, of a field theory that it 'approach[es] more closely to the truth' than an older matter-and-force theory should mean, unless words are being oddly used, that the ultimate constituents of nature are more like fields than like matter and force. But in this ontological context it is far from clear how the phrase 'more like' is to be applied. Comparisons of historical theories give no sense that their ontologies are approaching a limit: in some fundamental ways Einstein's general relativity resembles Aristotle's physics more than Newton's".[2] In the case of great scientific changes, we cannot take it for granted that language and experience are shared by the theories that succeed each other in time. As Feyerabend said, in these cases "it might be extremely difficult to compare successive theories in the usual manner, that is, by an examination of consequence classes", in order to maintain that a theory T_2 is closer or farther from truth than a theory T_1.[3] Revolutionary scientific changes raise the problem of incommensurability and they can induce the historian of science "to exclaim that when paradigms change, the world itself changes with them".[4]

For Kuhn, the idea of an increase in the verisimilitude of our theories can be applied only in the case of normal science, in which

[1] Quine 1970: 24.
[2] Kuhn 1987b: 265.
[3] Feyerabend 1987: 219f.
[4] Kuhn 1996a: 111.

change takes place within an accepted and shared disciplinary matrix. By talking about approximation to truth in an absolute sense, Popper made the mistake of attributing "characteristics of everyday research" to what happens in great revolutionary changes, in which it is not possible to discern a progress towards truth.[1] In Kuhn's view, the word 'truth', like the word 'proof', is "a term with only intra-theoretic applications":[2] "[t]here is [...] no theory-independent way to reconstruct phrases like 'really there'; the notion of a match between the ontology of a theory and its 'real' counterpart in nature [... is] illusive in principle".

This bias in favour of radical relativism is partly corrected by the fact that Kuhn preserves a certain idea of progressiveness in science. Of course, he denies that we can speak of scientific progress in teleological terms: we cannot conceive scientific progress as a march towards the discovery of the truth about nature. However, Kuhn believes it is legitimate to speak about scientific progress in a naturalistic and Darwinian sense, as the passage from less complex stages to more complex ones, which are characterised, for example, by greater "accuracy of prediction, particularly of quantitative prediction", by greater specialisation, by the greater "number of different problems solved". Accordingly, scientific development is "unidirectional and irreversible"[3] in the same sense as biological evolution, and for external observers it would be fairly easy to point out which are the older theories and which are the derived ones. This allows us to take a certain distance from the relativist thesis: we must acknowledge that not all scientific theories can be considered as of the same value for doing what scientists normally do.[4]

3. EMPIRICAL SUCCESS AND TRUTH. CRITIQUE OF CONVERGENT REALISM

A. Convergent realism and inference to the best explanation. Realism based on predictive success

In the more recent literature in the philosophy of science, the attack on verisimilitude-based realism is intertwined with the attempt to defend realism without using a notion the application of which requires a sort of 'computation', be it even conjectural, of the contents of truth and falsity of theories. Laudan – one of the severest critics of this attempt – has called it 'convergent epistemological realism', in order to distinguish it both from 'semantic realism' ("the claim that all theories have truth values and that some theories are true, although we know not which"), and from 'intentional realism' ("the view that theories are generally intended by their proponents to assert the existence of entities

---

[1] Kuhn 1987a: 19
[2] Kuhn 1987b: 266
[3] Kuhn 1996b: 206.
[4] See Kuhn 1996a: 170 ff; 1996b: 205f

corresponding to the terms in those theories").[1] Among the philosophers of science who have variously elaborated this view, there are Shimony, Boyd and Putnam – to be more precise, Putnam before his passage from metaphysical, or external, realism to empirical, or internal, realism.

At the basis of this defence of realism there is an argument generally known as 'Inference to the Best Explanation'. This argument is intended to support the idea that our best beliefs (for example, those codified in current scientific theories) are *closer to truth* than others (for example, past ones). 'Inference to the Best Explanation' sees progressive approximation to truth as the only empirical hypothesis capable of best explaining the functioning of so-called 'mature sciences', in particular their predictive successes. According to convergent realism, the only plausible and coherent explanation of the mature sciences' capacity to predict observable phenomena is to assume that the theoretical terms of successful theories refer to existing entities, and that these theories make at least some approximately correct statements about such entities. I believe this reasoning recalls Poincaré's argument against Le Roy's instrumentalist nominalism: science could not be a reliable rule or guide for action if it did not have also an objective value.[2]

According to the 'externalist Putnam', the various forms of 'anti-realism', including the positivist ones, have the serious defect of making "the success of science a *miracle*". The realist's allegation is that anti-realists are forced to leave the predictive success of theories unexplained. It is difficult to understand how "'electron calculi' and 'space-time calculi' and 'DNA calculi' correctly predict observable phenomena if, in reality, there are no electrons, no curved space-time, and no DNA molecules. If there are such things, then a natural explanation of the success of these theories is that they are *partially true accounts* of how they behave. And a natural account of the way in which scientific theories succeeded each other – say, the way in which Einstein's Relativity succeeded Newton's Universal Gravitation – is that a partially correct / partially incorrect account of a theoretical object – say, the gravitational field, or the metric structure of space-time, or both – is replaced by a *better* account of the same object or objects. But if these objects don't really exist at all, then it is a *miracle* that a theory which speaks of gravitational action at a distance successfully predicts phenomena; it is a *miracle* that a theory which speaks of curved space-time successfully predicts phenomena; and the fact that the laws of the former theory are derivable 'in the limit' from the laws of the latter theory has no methodological significance".[3]

It is clear from this observation on the passage from Newtonian to relativistic physics that Putnam, unlike other advocates of convergent realism, requires a certain continuity in great scientific transformations.

---

[1] Laudan 1984: 219.

[2] Poincaré 1946c, p. III, ch. X, sec. 2, 'Science, rule of action' (pp. 323-5).

[3] Putnam 1979: 18f.

If the development of science were marked by radical ruptures, and if it were not possible to detect some tendency towards the stabilisation of our ontological descriptions, so as to assign referents "to past theories from the viewpoints of their successors, and to recover past theories as limiting cases of their successors, [...] then realism would not be a tenable position".

In order to overcome this difficulty, Putnam supplied his realism with an appeal to the causal theory of reference and to the 'Principle of the Benefit of Doubt' (analogous to the principle of charity). It is thus possible to weaken the continuist requirement without giving it up completely. "By denying that the descriptions used in fixing the reference of scientific terms are synonymous with them", the principle of the benefit of doubt allows us to maintain a certain referential continuity across the change of the theories which describe the entities the terms refer to. It is sufficient to credit any scientific innovator (Newton, Einstein, Planck, Bohr, etc.) with referring to the entities posited by current theory (whichever they may be), provided they resemble sufficiently, in certain respects which have remained unvaried, the "descriptions he used so that it is reasonable to suppose he would have identified" these entities as the referents of his terms, "had he known more of the relevant facts". A weakening of the continuist requirement has been suggested also by McMullin, according to whom formal incompatibilities between theories can leave open the possibility of a certain continuity at the metaphorical level.[1]

B. Objections to convergent realism

This defence of realism is not unattractive. However, despite its ingenuity, in the last fifteen years many stringent objections have been raised against it. The most convincing criticisms revolve around two fundamental problems.

1. (Dis)continuity in scientific change – The first is historical. It concerns the high rate of discontinuity in scientific change, which makes the notion of convergence descriptively very problematic. It is enough to recall Kuhn's passages cited above in order to realise the difficulty of recognising any referential continuity, be it even charitably interpreted, in great scientific transformations. As Leplin, himself a realist, concluded disconsolately, "history appears to discredit our ability to identify the actual constituents of the world as much as it does our ability to learn their true natures".[2]

It must also be noted that, even if there were continuity, this would bring grist not only to the realist's mill. Boyd sees clearly that "[t]he two chief rivals of scientific realism, empiricism and constructivism, are each capable of providing plausible explanations for the apparent semantic and methodological continuity in the history of well-developed and mature sciences. Indeed, they offer variations on the

---

[1] Leplin 1984: 5f, also with reference to McMullin 1984.

[2] Leplin 1984: 2.

same explanation: the continuity in question is a manifestation of linguistic, conceptual, and methodological *conventions*".[1] Boyd rightly numbers both Carnap and Kuhn among the constructivists.[2] We could certainly add Reichenbach and, especially, Cassirer. As I pointed out in the previous chapter, the idea of considering scientific change to be a sort of evolutionary process in which the previous stages are somehow preserved in the subsequent ones is not foreign even to logical idealism: the possible convergence towards a stable theoretical arrangement of concepts and principles could be interpreted as the progressive uncovering of the "ultimate logical invariants" of experience at the basis of scientific knowledge and its development.

2. Empirical success warrants truth? – The second kind of criticisms concerns the possibility of establishing a connection between the empirical success of a theory and its truth or degree of verisimilitude: since the same complex of phenomena can be explained by means of a plurality of alternative theories, we cannot take for granted the connection we are talking about. Laudan formulated the best known and most cogent objections: in 'A Confutation of Convergent Realism', he tried to show, more than the falsity of metaphysical realism, the inadequacy of the argument based on the idea that realism is the best empirical explanation of the predictive successes of science – the explanation that does not make those successes "a miracle". Laudan's objections are both (1) factual and (2) logical, as befits the critique of a thesis which claims to be an empirical explicative hypothesis.

(1) On the *factual plane*, Laudan cites quite an impressive number of cases from the history of the mature sciences, in order to highlight two aspects of scientific activity which do not accord with convergent realism.

(a) There are no historically grounded reasons to think that the protagonists of scientific change act (or *must* act) according to the retentionist "strategy of seeking to preserve" the mechanisms, models and laws of "earlier theories in later ones".[3] Nor is there historical backing for the more modest requirement "that every satisfactory new theory must be able to explain why its predecessor was successful insofar as it was successful". If a theory is better supported than its rivals and its predecessors, it is not epistemically decisive that it explains why the previous theories worked: "a theory's ability to explain why a rival is successful is neither a necessary nor a sufficient condition for saying that it is better than its rival".[4]

(b) Laudan undermines the foundations of the convergent realist's thesis of the interdependence between the property of empirical success and the properties of referential success, truth and verisimilitude. This interdependence must be considered very problematic for two

---

[1] Boyd 1984: 79.
[2] Boyd 1984: 55-8 ('Empiricism and Constructivism').
[3] Laudan 1984: 235.
[4] Laudan 1984: 240f.

simple historically documentable reasons: (i) there are theories which, according to charitably applied contemporary standards, have referential terms, are true or approximately true, but are not empirically successful (for example, at its beginnings, atomic chemistry); (ii) there are empirically successful theories which neither have referential terms, nor are true or approximately true (particularly convincing examples are the phlogiston theory, the theory of the caloric, and, especially, the theory of luminiferous aether, which – according to Maxwell – "was better confirmed than any other theoretical entity in natural philosophy".)[1] To sum up, historical evidence does not seem to back the convergent realist's thesis, that referentiality, truth, or even only approximate truth, are the only possible explanations of the empirical success of science.

(c) We can add to the factual criticisms that the history of science does not even support unequivocally the thesis of realism's methodological superiority over the anti-realist positions (in particular over positivism) from the point of view of scientific development. This idea has acquired a certain strength, especially after the experimental proofs of the existence of atoms, considered by the realists as a clear refutation of Mach's sensationalism.[2] But, as Arthur Fine has shown, many characteristics of the development of recent theories, such as relativity and quantum theory, do not fit in with the conviction that realism fosters scientific progress more than anti-realism (it is sufficient to think of the operationalist and phenomenalist tendencies clearly present in many aspects of Einstein's, Bohr's and Heisenberg's work).[3]

(2) To these factual objections, based on the history of science, Laudan added others concerning the *logical strength* of the 'Inference to the Best Explanation' as it is used to defend the realist conception of science and scientific progress.

(a) The first objection exposes the lack – which I already noted in the discussion of Popper – of a formally adequate semantical notion of approximation to truth. Contrary to what some advocates of realism claim, the convergent realist is not justified in being content with the intuitive content of the concept of verisimilitude: "[i]f the realist would demystify the 'miraculousness' (Putnam) or the 'mysteriousness' (Newton-Smith) of the success of science, he needs more than a promissory note that somehow, someday, someone will show that approximately true theories must be successful theories". In the absence of a rigorous analysis, "the realist seems to be long on intuitions and short on either a semantics or an epistemology of approximate truth".[4]

(b) But the worst problem is the logical weakness of the argument with which the convergent realist tries "to show the skeptic that

---

[1] Laudan 1984: 226.

[2] For Mach's sensationalism and for his relationship with twentieth century physics, see Wolters 1989, 1990.

[3] See Fine 1984: 91ff

[4] Laudan 1984: 230 (for the references to Newton-Smith, see 1984: 246f, nn. 14, 15, 17, 19).

theories are not ill gotten, the positivist that theories are not reducible to their observational consequences, and the pragmatist that classical epistemic categories (e.g., 'truth' and 'falsehood') are a relevant part of metascientific discourse".

The point is that the realist makes use of "an abductive inference which proceeds from the success of science to the conclusion that science is approximately true, verisimilar, or referential (or any combination of these)". But, in so arguing, the realist forgets that "ever since antiquity critics of epistemic realism have based their skepticism upon a deep-rooted conviction that the fallacy of affirming the consequent is indeed fallacious". From Sextus Empiricus to Hume, "many nonrealists have been nonrealists precisely because they believed that false theories, as well as true ones, could have true consequences".

Putnam, Boyd and Newton-Smith's claim that "epistemic realism can reasonably be presumed true by virtue of the fact that it has true consequences" looks like "a monumental case of begging the question": if "[t]he nonrealist refuses to admit that a *scientific* theory can be warrantedly judged to be true simply because it has some true consequences", it is highly unlikely that he will "be impressed by the claim that a philosophical theory such as realism can be warranted as true because it arguably has some true consequences. If nonrealists are chary about first-order abductions to avowedly true conclusions, they are not likely to be impressed by second-order abductions, particularly when [...] the premises and conclusions" are based upon conceptual and factual notions and presuppositions which are so imprecise, unwarranted, and in some respects presumably false.[1]

In order to by-pass the objection, we could say perhaps that "the realist is not out to convert the intransigent skeptic or the determined instrumentalist [...]. Perhaps, he is seeking to show that realism can be tested like any other scientific hypothesis, and that realism is at least as well confirmed as some of our best scientific theories".

Prima facie, this way of putting things is quite plausible and attractive. However, a deeper analysis exposes the fact that the answer is not in line with one of the fundamental requirements of the methodology of epistemological realism. It is precisely the realists who oppose *ad hoc* and *post hoc* theories, maintaining that an hypothesis cannot be considered reasonably true, or well confirmed, "just on the strength of the fact that its thus-far-tested consequences are true. [...] Before a realist accepts a scientific hypothesis, he generally wants to know whether it has explained or predicted more than it was devised to explain, whether it has been subjected to a battery of controlled tests, whether it has successfully made novel predictions, and whether there is independent evidence for it".[2] Therefore, even granting for argument's sake realism's capacity to explain the success of science, for the present, realism is a mere *ad hoc* hypothesis, which has not yet led to new predictions, and

[1] Laudan 1984. 242f.
[2] Laudan 1984 243

which is not confirmed by other data beyond those which are already known, and for the explanation of which it was expressly introduced. To sum up, if realism is to be evaluated as any other scientific explanatory theory, the realist's methodological canons themselves will condemn it as a not particularly valuable hypothesis.

(c) An analogous criticism of the logical strength of the 'best explanation' argument has been formulated by Arthur Fine. He has acutely observed that the convergent realist's reasoning resembles a kind of metatheoretical argument intended to show the validity of the theoretical inferences by means of which science posits certain entities and gives certain descriptions of reality.

But we have known since Hilbert's time that metatheoretical arguments should be "more stringent" and secure than theoretical ones. Since the problem raised by the anti-realists is whether acceptable and good explanatory hypotheses should be considered true precisely because they are acceptable and good, one cannot defend realism by saying that it is an acceptable and good explanatory hypothesis about the success of science.

On the other hand, in order to defend this thesis with normal inductive arguments, one should be able to find a considerable number of inductions capable of connecting the observations of the predictive successes of science with the referentiality and the at least approximate truth of science's posits and descriptions. But we are by now well aware of the impossibility of observing a connection of this kind, since we are not able to say what reality and truth look like independently of our theories. Therefore, we do not possess observations which could constitute the starting point for an inductive inference. We are again faced with the "well-known idea that realism commits one to an unverifiable correspondence with the world".[1] As I have repeatedly stressed, this very idea is the main argument at the basis of sceptical doubts and of Kant's analysis of the object of representations, which I considered in the previous chapter.

4. THE DRAW BETWEEN METAPHYSICAL REALISM AND ANTI-REALISM

4.1. RUSSELL VS POINCARÉ AND THE VIGOUR OF REALISM

However, these objections, despite their strength, do not seem to me to show metaphysical realism's logical inconsistency or total emptiness, not to mention its meaninglessness. In recent years, many philosophers of science and philosophers of language have continued to uphold forms of metaphysical realism, justifying this choice by means of arguments connected with the causal theory of reference and with the conception of names as rigid designators.

[1] Fine 1984: 85f.

Lately, Ian Hacking has formulated a peculiar kind of scientific realism by connecting these topics with methodological considerations on the nature of scientific experiments and on the relationships between theorisation and experimentation. Referring to the experimental and technological dimension of science – long neglected by philosophers of science – Hacking tries to curb the more drastic versions of the thesis of the incommensurability of theories, and to secure a high degree of objectivity for the scientific enterprise. His position is founded on the distinction between 'realism about theories' and 'realism about entities'. Hacking suggests we give up the conception – now deemed indefensible – of theoretical change as a more or less continuous approximation to the truth about nature. At the same time, however, he keeps firm the real existence of the non-observable entities hypothesised by scientific theories; more precisely, he claims the existence of those entities which can be manipulated in order to intervene causally in the course of natural events, or in order to produce phenomena in laboratory experimentation.

If we restrict ourselves purely to the theory of elementary particles, it is very difficult, not to say impossible, to present conclusive proofs of the existence, for instance, of electrons. But, Hacking says, we are persuaded of their reality the moment we abandon the dualism of knowing and doing (already criticised by Dewey and by the young Marx), and we pay due attention to experimental devices, such as the polarising electron gun *Peggy II* that 'sprays electrons', allowing us to interfere causally with other more hypothetical parts of nature. In short, the best proofs in favour of scientific realism of entities seem to come from engineering (the locus of intervening and doing), and not from theorisation (the locus of representation and thought) .

On this basis, Hacking criticises explicitly the central thesis of the internal realism Putnam adopted in the place of external realism. According to internal realism, the question "*what objects does the world consist of?* is a question that it only makes sense to ask *within* a theory or description. 'Truth', in an internalist view, is some sort of (idealized) rational acceptability – some sort of ideal coherence of our beliefs with each other and with our experiences *as those experiences are themselves represented in our belief system*". At the metaphorical level, Hacking is willing to share the statement that *we* "cut up the world into objects when we introduce one or another scheme of description". But he also declares that he cannot endorse the previous statement that "'Objects' do not exist independently of conceptual schemes".[1]

The "Inuit are said to distinguish ever so many kinds of snow that look pretty much the same to us", by means of which "[t]hey cut up the frozen North by introducing a scheme of description". But, according to Hacking, "[i]t in no way follows that there are not 22 distinct mind-independent kinds of snow, precisely those distinguished by the Inuit". Furthermore, since the Inuit do not ski, it can very well be the

[1] Putnam 1982: 49f, 52, cited in Hacking 1983: 93.

case that "powder snow, corn snow, or Sierra cement spoken of by some skiers neither contain nor are contained in any Inuit class of snow". Nevertheless, we are allowed to presume that "there is still powder snow *and* all the Inuit kinds of snow", and that such kinds all refer to "real mind-independent distinctions in a real world." "These remarks – says Hacking – do not prove that there is powder snow, whether anyone thinks of it or not. They merely observe that the fact that we cut up the world into various possibly incommensurable categories does not in itself imply that all such categories are mind-dependent".[1]

The vigour of realism lies precisely in the undeniable possibility of using our linguistic expressions in order to refer directly to the object, its existence and its properties, independently of the possible conditionings by the linguistic, theoretical and methodological conceptual schemes within which we operate, and sometimes are forced to operate in the absence of conceivable alternatives. We can change the way we call things as we wish, but they will not change their properties because of this. As the poet says, expressing this idea with a vividness superior to all philosophical subtleties, "[t]hat which we call a rose / By any other name would smell as sweet". Although we come across this quotation very often, it is seldom recalled that to give up the loathed name of Montague does not only mean not to give up any of the parts which make up Romeo's body, but it does not even entail losing the "dear perfection" that emanates from their conjunction – that perfection which, in Duhem's view, is one of the reasons that induce us to lend ontological value to our theoretical constructions.

Juliet's passionate words can be invoked to deny ontological relevance not only to the conventionality of linguistic symbols, but also to all epistemic conditions: we can always act on a so-called 'objectocentric' expressive plane, saying that the object is what it is, or has or does not have, certain properties, independently of the linguistic, theoretical and methodological presuppositions on which our knowledge of the object depends. As a matter of fact, those very presuppositions will have to be judged correct or incorrect according to whether they are or are not consistent with the object.

This was exactly the point underlying Russell's assertion of the objective value of spatial congruence and his criticism of Poincaré's geometrical conventionalism. "It seems to be thought that, since measurement is required to *discover* equality and inequality, there cannot *be* equality or inequality without measurement. The true inference is exactly the opposite. What can be discovered by an operation must exist apart from the operation: America existed before Christopher Columbus, and two quantities of the same species must *be* equal or unequal before measurement. Any method of measurement is right or wrong according as it brings out the right or wrong result. M. Poincaré,

[1] Hacking 1983: 95; see also Hacking 1984.

on the contrary, holds that measurement *creates* equality and inequality".[1]

4.2. BOLZANO AND THE DIFFICULTIES OF REALIST ESSENTIALISM

A natural development of this realist point of view is the full development of the consequences of Bishop Butler's famous remark (today very fashionable again) that "everything is what it is and not another thing". Since a thing (interpreted as an individual, a substance, or a natural kind) is what it is in virtue of its having certain properties and not others, it follows that if there are some properties which make an object what it is, these will be essential properties of the object as distinct from its purely accidental ones.[2]

A form of the old Aristotelian essentialism has thus come to the forefront again, refurbished by means of a not always clear and sound mixture of abstract philosophical arguments, and logical and linguistic analyses connected with modal logic and with the functioning of proper names, singular terms and descriptions. New and surprising conceptual monsters have been born – such as the contingent a priori and the a posteriori necessary – from the cross between that dinosaur of philosophical speculation which seemed definitely extinct (Aristotelian essentialism), and animals which have always thrived within the analytic zoo, such as Leibniz's idea of possible worlds and the notions of analyticity, apriority, etc.

The acknowledgment of the fact that similar speculative constructions cannot be dismissed as meaningless must not make us forget the connection between realism and the sceptical stance which I have repeatedly emphasised. *If* we intend those constructions to characterise an essentialist metaphysical realism, we can oppose against them the Kantian epistemological considerations which I made on the nature of the object of knowledge, not to mention the logical and semantical arguments, connected with Löwenheim-Skolem's theorem, which determined Putnam's passage from external to internal realism, and on which discussion is currently thriving.[3]

---

[1] Russell 1990: 396f.

[2] The issue was discussed also by Campbell with very interesting arguments (see Campbell 1957: 47-9 ('Defining and non-defining properties')).

[3] See Putnam 1983b. Following some remarks by Paolo Leonardi and Stephen Barker, I would like to specify that I am merely interested in maintaining that Kripke's ideas *can* be used for essentialism. I am not attempting an exegesis of Kripke's genuine position – nor do I enter into the question whether the doctrine of rigid designations and a posteriori necessary truths presupposes essentialism (see Parrini 199+b). I leave all these questions to metaphysical realists: it is up to them to decide whether metaphysical realism implies essentialism. I limit myself to suggesting that metaphysical realism, as well as the distinction between essential and inessential properties of things in themselves, should be considered *devoid of cognitive value*.

I was surprised to learn recently that these convictions were shared by a thinker who was not particularly indulgent of subjectivist conceptions, namely Bernard Bolzano, who in the *Wissenschaftslehre* gave a particularly incisive formulation of the difficulties threatening the attempt to distinguish between essential and inessential attributes of objects metaphysically not considering our ways of referring to the objects.

Bolzano opts for a definition of the distinction between essential and inessential qualities which connects this distinction to the concepts used to speak about things in each particular occasion. "According to [that] definition, we cannot determine from a given attributive idea alone whether it is an essential or inessential attribute of some object. Rather, this depends in part upon the object whose attribute it is supposed to represent, and in part upon the concept that we form of this object. The same attribute will sometimes be called essential and sometimes inessential of the same object, depending upon the concept under which we susbsume the object".

One might object against such a definition that "the basis for my decision whether or not a certain attribute is essential of an object, namely the concept under which we subsume the object, is altogether too arbitrary and insignificant a circumstance". Nevertheless – Bolzano notes – "if we consider that no object is known to us save through the idea under which we have subsumed it, and that it cannot become an object for our enquiry except through this idea, then we must admit that for the purposes of logic it is neither an insignificant nor an arbitrary circumstance whether a given object is subsumed under one idea or another. As an object of our *knowledge* a thing is no more than what we represent in our minds, whenever we believe that we represent *it*. Thus, for the purposes of logic, its idea constitutes [exhausts] its *essence*".[1]

In these words there is perhaps something more than an echo of Kant's analysis of the object of representation cited in Chapter IV. Here too the aim is to establish the impossibility of speaking, if not of the object, certainly of our knowledge of it, except in relation to our representations. This raises again the problem of the dependence of our statements about the existence and the properties of things upon a conceptual system.

As a matter of fact, if we look at the choice between realism and idealism from the point of view of the most recent discussions, it seems fair to conclude that, at least so far, no decisive arguments have been offered either in favour of forms of metaphysical realism or in favour of various forms of anti-realism (idealism, phenomenalism, instrumentalism, and so on). Both conceptions have advantages and defects which somehow balance each other out.

Anti-realists have rightly reminded realists: (a) that our knowledge requires subjective epistemic assumptions; (b) that what we know as object, and call 'object', is in fact the result of a synthesis of experience

[1] Bolzano 1972. 159f [I, 520f, 522].

operated on the basis of such assumptions; and, lastly, (c) that there is an apparently indissoluble tie between scepticism and realism.

On their side, realists have been able to reply to anti-realists by stressing: (a) that the absolute notion of object is neither meaningless nor internally incoherent; (b) that there are no arguments to establish that our epistemic conditions of knowledge, uncovered by the analysis of the cognitive process, could not as well be structural properties of objects in themselves; and, lastly, (c) that the development of knowledge depends on a given which is not completely determinable by the knowing subject, that is, by the epistemic structures of reference – and this dependence suggests quite naturally the idea of a reality independent of knowledge and which knowledge itself strives to reach.

This undeniable presence of strong and weak points in both metaphysical positions prompts the question whether there could be a different way of looking at the whole question, a way that aspires to eliminate the problem precisely in so far as it is a metaphysical problem. The neopositivists (in particular Carnap) tried something of this kind by means of the verifiability principle. Something similar has surfaced again in the recent debate on scientific realism.

5. THE SCOPE OF THE SCIENTIFIC METHOD AND NON-METAPHYSICAL REALISM

5.1. THE NATURAL ONTOLOGICAL ATTITUDE

In a 1984 essay, entitled 'The Natural Ontological Attitude', Arthur Fine proclaimed the death of realist metaphysics. He proposed in its stead a "minimalist" realism, suitable for our "postrealist" times characterised by the predominance of peculiar realisms – such as Horwich's semantical realism and Putnam's internal realism – and peculiar forms of anti-realism, such as Rorty's "epistemological behaviourism". According to Fine, the watershed between the realist and the post-realist age is the "natural ontological attitude".

Fine maintains that scientific theories make genuine existential statements (semantic realism) and he considers as wholly legitimate the inferences which lead to acceptance of scientific hypotheses and theories based on certain empirical proofs. Therefore, he is willing to admit that the predictive success of a theory constitutes a proof of the referentiality of its central terms, irrespective of whether these refer to observable or non-observable entities.

However, Fine distances himself from traditional metaphysical realism by assuming the "natural ontological attitude", according to which the "acceptance of a hypothesis need not [...] involve any of the metaphysical implications which the truth of the hypothesis has been thought to carry in the realist tradition". On the basis of the 'natural ontological attitude', "the explanatory and inferential structures of scientific reasoning" are autonomous, and the "metaphysical interpretation is a dispensable superaddition".[1] According to me, in a minimalist perspective, one looks at the empirical proofs of the hypothesis that there are electrons as empirical proofs that there are electrons, without considering the hypothesis that 'really' there are electrons, where 'really' is interpreted in the sense that the realist metaphysician has at heart.

It does not seem possible to give a content to the metaphysical formulation by passing from the first-order abduction that there are electrons to the second order abduction that there really are electrons. The first abduction is a legitimate inference, which can undergo all the procedures of strengthening and weakening to which it is possible to subject scientific inferences, according to the more or less strict canons fixed by the various theories of methodology. The second is a completely gratuitous inference, which cannot be verified, especially if we require controlled tests in accordance with the realist's particularly rigid methodological requirements.

This induces Fine to trace back the realist metaphysical position to motivations and aspirations of a religious nature: "[i]n support of realism there seem to be only those 'reasons of the heart' which, as Pascal says, reason does not know. [...] [B]elief in realism involves a profound leap of faith, not at all dissimilar from the faith that animates deep religious convictions". We may enjoy the "intricate and sometimes beautiful philosophical constructions" elaborated by metaphysical realists by intertwining the notions of meaning, knowledge, reference, truth and so on; but we cannot but consider vain their claim to ground their faith rationally and to transform their conceptual edifices into something more than "wonderful castles in the air".[2]

To sum up, Fine is willing to accept the inferences and the scientific arguments with which we back up the (provisional) truth of certain hypotheses and theories. Consequently, he is also willing to accept the existence of the entities posited by such hypotheses and theories. But he denies "that philosophical accounts of truth and reference" have any "explanatory role in understanding scientific success or underwriting scientific conclusions. Science all by itself says all that needs to be said or can defensibly be said in response to philosophical questions about the nature and status of scientific knowledge".[3]

5.2. CARNAP'S EMPIRICAL REALISM

In my view, the general philosophical teaching of the recent epistemological debate about realism is the following. The verdict on metaphysically interpreted epistemological oppositions transcends the conditions of decision proper to the normal scientific procedure: it goes beyond

---

[1] Leplin 1984: 7.
[2] Fine 1984: 104, n. 7.
[3] Leplin 1984: 7; see Fine 1984: 97ff.

the decision methods normally employed in scientific inquiry. This is exactly what the logical empiricists (especially Carnap) maintained, when they numbered epistemological-metaphysical controversies among pseudo-problems. They declared them *meaningless* in the name of a general conception of meaning – the verifiability principle – which was subsequently exposed as highly problematic (it should be noted however that in the *Aufbau* and in 'Scheinprobleme in der Philosophie', Carnap tended to use 'devoid of meaning', 'devoid of cognitive value' and 'devoid of scientific value' as essentially equivalent expressions).[1] But the crisis of the logical empiricists' conception of meaning does not affect the kernel of truth present in their formulations – a truth which is making itself heard again in post-neopositivist philosophy of science.

The solution I am referring to was elaborated by Carnap in the course of his whole philosophical activity. It rests on the distinction between two notions of reality (metaphysical and empirical) and between two corresponding notions of realism (metaphysical and empirical). Carnap's proposal dates from the end of the twenties, but it was subsequently incorporated in the dichotomy of the fifties between external and internal questions of existence.

As I have tried to show elsewhere,[2] this distinction allowed Carnap to accept realism based upon the empirical concept of reality: this realism was held to be verifiable, and therefore meaningful. At the same time, this distinction enabled Carnap to reject all forms of realism, idealism, phenomenalism, spiritualism, materialism and instrumentalism, qua metaphysical positions devoid of empirical and cognitive meaning. In other words, the distinction between empirical and metaphysical reality makes it possible to reject all '*ontological*' commitment in the properly metaphysical (and therefore unverifiable and meaningless) sense of the expression, without being compelled by this to reject any ontological commitment in the empirical sense – a commitment that we could call *ontic* and that is very close to Fine's natural ontological attitude.[3]

The notion of *empirical reality* (also believed to be verifiable, and therefore endowed with meaning and with scientific and cognitive value) is for Carnap the notion we apply in all the contexts in which we distinguish between reality on the one hand and dream, hallucination, legend and so on, on the other. All scientists agree on this notion, whatever their metaphysical creed, realist or idealist, spiritualist or materialist. "It is this concept which distinguishes a geographically determined mountain from a legendary or a dreamed mountain, and an experienced emotion from a simulated one".[4] The 'construction', or, better, the 'constitution', of the physical, psychological and cultural objects belonging to empirical reality, as distinct from those belonging to non-reality (dream, hallucination, legend, invention, hypnotic suggestion,

---

[1] See Parrini 1994b: 261f.

[2] See Parrini 1994b.

[3] See Parrini 1994b: 260, 273f.

[4] Carnap 1967a: 273.

etc.) depends: (1) on the possibility of including 'real' objects in a unitary whole of entities placed at least in the temporal order and ruled by laws in accordance with the principles or rules of the *Konstitutionssystem*; and (2) on partly empirical and partly pragmatic considerations of balance regarding the entirety of our cognitive claims, i.e. – as Quine was to say some years later – on "considerations of equilibrium affecting the field [of our beliefs] as a whole".[1]

This concept of empirical reality is different from that of *metaphysical reality* (held to be unverifiable and therefore devoid of meaning and of scientific and cognitive value). We run into it when we leave the domain of empirical research and venture into the debates among the various kinds of realists, phenomenalists and idealists – when, for example, we wonder whether an object which has empirical reality has also another peculiar kind of reality, in a special sense of this word. Carnap specifies this sense in general terms, sometimes by means of the idea of transcendence, sometimes by means of that of 'independence from the cognising consciousness'. He is particularly clear in some pages of 'Scheinprobleme in der Philosophie' and of the 'Replies and Systematic Expositions', where metaphysical reality is implicitly identified with absolute reality, or reality in itself.[2] It is on the basis of arguments of this kind that Schlick too at a certain point abandoned the metaphysical realism of the *Allgemeine Erkenntnislehre* in favour of the specifically neopositivist theses of 'Positivismus und Realismus'.

Therefore, the distinction between two notions of reality makes possible an alternative answer to the traditional controversy between idealism and realism – an answer which consigns the controversy to the domain of pseudo-problems devoid of empirical value. This distinction allows us to do the same with the opposition between instrumentalism and scientific realism. If we look at this opposition keeping in mind this distinction, we obtain two pairs of possible philosophical positions which are each in contrast with the other.

(1) We have the *first pair* if we look at the debate between scientific realists and instrumentalists from the point of view of *metaphysical realism*. (a) The instrumentalist denies the *metaphysical reality* of theoretical entities, though he is willing to allow that of observational entities. (b) The scientific realist upholds the *metaphysical reality* not only of observational entities, but also of theoretical ones, by virtue of the homogeneity and continuity that exists, according to the scientific realist, between these two kinds of entities.

(2) We have the *second pair* of positions when we look at the debate between scientific realists and instrumentalists from the point of view of *empirical realism*. (a) The instrumentalist denies the *empirical reality* of theoretical entities, considered only 'useful fictions', but is willing to allow that of observable entities. (b) The scientific realist upholds the *empirical reality* both of observable and theoretical entities,

---

[1] Quine 1964d: 43; see also p. 42.
[2] See Carnap 1967a: 273, 275, 281, 333f; 1963b: 873.

because the differences between them are not, in the scientific realist's view, great enough to justify a different ontic attitude.

It is possible to choose the alternative that recognises the empirical reality of all entities (observable or not observable) posited by science with good reasons, while rejecting as devoid of empirical value the question of the metaphysical reality and truth of the conceptual network, based on the ontological dualism of subjects and objects (both observable and not observable) with which science operates.

6. RELATIVISM, SCIENTISM, AND ANTI-METAPHYSICS

Therefore, in the light of sections 5.1. and 5.2., it seems possible to assume a position on the controversy between realists and anti-realists (phenomenalists, idealists, etc.) – and in particular between instrumentalists and scientific realists – which is analogous both to Fine's natural ontological attitude, and to Carnap's position (except for its semantical import). However, there remains the problem of understanding more precisely the sense in which the metaphysical outlook is rejected when we are no longer willing to accept the verifiability principle and to declare metaphysics meaningless. In my view, it will be possible to reject metaphysics in the sense which has come to light in the recent debate on realism. I tried to show elsewhere that this rejection of metaphysics expresses very clearly the fundamental motivation of the neopositivist, and especially Carnap's, (not always well formulated) anti-metaphysical stance. Metaphysics can be rejected by saying that it raises a problem of an absolute kind, transcending in principle what science can establish. Therefore, metaphysics raises a scientifically spurious question, a question *devoid of scientific value*.

6.1. QUINE AND CARNAP ON ONTOLOGY

I will clarify this by recalling the controversy on ontology between Quine and Carnap in which many thinkers saw the final defeat of the anti-metaphysical programme of the neopositivists in general, and of Carnap in particular.

Since the fifties, Quine has declared that "[t]he fundamental-seeming philosophical question, How much of our science is merely contributed by language and how much is a genuine reflection of reality? is perhaps a spurious question which itself arises wholly from a certain particular type of language. Certainly we are in a predicament if we try to answer the question; for to answer the question we must talk about the world as well as about language, and to talk about the world we must already impose upon the world some conceptual scheme peculiar to our own special language".[1]

[1] Quine 1964c: 78.

After the crisis of the verifiability principle, it seems to me difficult to find a meaning for the word 'sense' which enables one to deem as 'senseless' the question of the absolute correctness of a conceptual scheme or, in Carnap's terms, of a linguistic form. But this does not entail that we have to renounce also a non-linguistic version of the anti-metaphysical, anti-absolutist, stance of positive philosophy. If we understand what it means to affirm the truth or falsehood of a sentence relative to a conceptual scheme, it seems to me obvious to suppose that we also understand what it means to affirm the absolute (non-relative) truth both of a sentence and of those sentences by which we can express the informing principles of a conceptual scheme.

But to recognise the meaningfulness of this statement does not mean to admit it belongs to the *Gebiet der Wissenschaft* [domain of science].[1] On the contrary, once we have acknowledged (also for reasons partly connected to the crisis of the verifiability principle) that science works within a conceptual system of assumptions, and makes its more or less radical and 'revolutionary' choices by constantly mixing pragmatic considerations with empirical reasons, we have implicitly accepted a characterisation of scientific questions which excludes from science the metaphysical problems concerning the *Ding an sich* and the absolute correctness of a conceptual scheme, an hypothesis or a theory, in the sense of their conformity to a reality in itself, independent of any conceptualisation. Quine may even be right in claiming that the "quest of ultimate categories" and the "limning of the most general traits of reality"[2] cannot in principle be distinguished from the working out of specific scientific hypotheses and theories; but none of these activities must be mistaken for an attempt to capture reality *an sich,* i.e., as used to be said, the 'essence' of things.

These considerations lead to conclusions on the controversy between scientific realism and instrumentalism, as well as on the general epistemological controversy between realists and anti-realists (phenomenalists and idealists of various kinds), analogous to those reached by Laudan's and Fine's criticisms of the methodological deficiencies of convergent realism. Unless we can resuscitate an epistemologically (not only logically) acceptable form of the verifiability principle, we will not be able to say that the metaphysical contrast between scientific realism and instrumentalism, between realism and anti-realism of various kinds, is *devoid of meaning*, or even only *devoid of cognitive value*. But we will certainly be able to say that it does not belong to the *domain of science*.

As we know it, science can only posit – more or less hypothetically, and chiefly by means of its techniques of experimental manipulation, rightly reappraised by Hacking – a certain domain of entities, giving us the descriptions of it which science considers most adequate, distinguishing between things which are real and things which are imagi-

---

[1] Carnap 1967a: xi.
[2] Quine 1970: 161.

nary, dreamed, etc.. But science has nothing to say about reality in itself, that is, about the absolute correctness of its posits and descriptions. In this typically positivist claim (firmly maintained by Hempel),[1] it seems that there is a kernel of truth – of positivist truth – which it is difficult to forsake, and which has really not been completely abandoned even after the crisis of neopositivism. This is proved by those of Laudan's and of Fine's arguments which show how the decision on metaphysically interpreted epistemological controversies transcends the conditions of decidability proper to normal scientific procedure. It is not possible to make second order abductions about the 'real' reality of objects, or about the 'true' truth of our hypotheses and theories, if the expressions 'real' and 'true' are interpreted in a metaphysical sense which transcends, or anyhow exceeds, the sense in which science attributes reality to certain entities and truth to certain hypotheses and theories.

6.2. ANTI-METAPHYSICS AS ANTI-ABSOLUTISM

Therefore, the neopositivist criticism of metaphysical absolutism seems to enjoy a relevance to the present and an intrinsic strength which seem definitely underestimated in many areas of contemporary philosophy. This is not to say that this criticism does not have weak points: in fact, it meets with a considerable problem because its rejection of metaphysics depends upon a relativist conception. The problem is that, if we adopt this approach, which is completely independent of the verifiability principle, it is possible to give philosophical value to anti-absolutism only if we take up two very problematic assumptions. Once we drop the verifiability principle, we will no longer be able to say – as Carnap and Schlick claimed to do – that metaphysical realism, in so far as it is founded on the idea of a transcendent reality in itself, is a thesis devoid of empirical content and therefore of meaning. Consequently, it will be possible to pass from the statement that questions concerning reality in itself do not have scientific value to the wider one that they do not have cognitive value, only if we assume the validity of: (1) a scientistic position which ascribes genuine cognitive value to science only; and (2) a relativist conception of truth and cognitive objectivity, which ascribes to them an exclusively intratheoretical or intrasystemic value.

Scientism is unattractive for various reasons, but it is radical relativism that presents the worst difficulties. Taken in large doses, relativism is a highly problematic philosophical position. It is very hard to give up any distinction between what our theories say, the reasons we have for believing them, and their truth and falsehood, in a sense of truth and falsehood not purely relative to a set of epistemic conditions of reference. A total renunciation of these distinctions could easily lead to a self-defeating position. The problem I will have to face next, in order to salvage a form of positive (if not positivist) philosophy, is that of

[1] See Hempel 1988: 14; 1992: 48, 51.

focusing a more general perspective, capable of warranting an objectivist component as significant as the previously secured relativism and empiricism. This will be the subject of the next chapter. In it, I will try to elaborate notions of objectivity and truth which, though not totally relativist, are not compromised with the absolute conception of the object and with the correspondentist view of truth proper of metaphysical realism.

Chapter VI

TRUTH, COGNITIVE OBJECTIVITY, METAPHYSICS

1. THE PROBLEMS OF RADICAL RELATIVISM

In the previous chapter, I argued that the crisis of the verifiability principle does not allow us to consider the metaphysical opposition between realism and the various forms of anti-realism as a meaningless problem. Yet, it is still possible to consider the controversies about the nature of knowledge as pseudo-scientific when they get involved in problems of an absolute kind that, as such, go beyond the sphere of decidability proper to science. Since scientific investigation is structurally connected with the use of a particular linguistic, theoretical, methodological and axiological framework, metaphysical questions which refer to the absolute can find no place in it.

Accordingly, we might be tempted to retain the validity of positive philosophy by attributing genuine cognitive value only to science (scientism), and by adopting at the same time an entirely intrasystemic, or intratheoretical (i.e., relativist), conception of truth and objectivity. All this would enable us to eliminate positivistically, as devoid of scientific and cognitive value, the question of the existence of a reality in itself, a transcendent reality, and of the correspondence of our cognitive claims to that reality.

However, I do not think that we can follow relativism right to the end, going beyond the moderate epistemic relativism I defended in Chapter II. A complete relativisation of truth (and objectivity) runs into two kinds of difficulty which, in my view, make relativism even less acceptable than pragmatism, (logical) idealism, and metaphysical realism.

1.1. OBJECTIVITY AND THE NORMATIVE CHARACTER OF TRUTH

The first difficulty consists in the fact that radical relativism ends up compromising in an irreparable and not easily defensible way those aspects of the notion of truth by which truth is seen as a unitary axiological and normative ideal. This point has always been emphasised not only by metaphysical realists, but also by philosophers of neo-Kantian, phenomenological and neopositivist orientations, who oppose a conception of truth as the reflection of a transcendent reality. This has been emphasised again in recent years, precisely in order to criticise the relativist outcomes of the new philosophy of science.[1]

---

[1] It should be clear in the end that the fact that one does not equate being true with being believed to be true does not – as is often assumed – commit one to metaphysical correspondentism, i e., to the thesis that a statement is true if and only if there is something in reality that makes it true.

It has thus been emphasised that one of the fundamental characteristics of science and other human activities is the value of objectivity. To pursue this ideal in the various fields of knowledge, as in everyday life, means to commit oneself to searching for, and to applying, criteria for the testing of our assertions which must be detached, impartial and independent. That is, our judgements are not to defer to subjective or personal authorities in the hands of single individuals, or more or less organised social groups.

An exclusively relativist approach clashes with this regulative value, in so far as it subordinates completely the validity of our assertions to the authority of a conceptual scheme which is not, in its turn, subject to critical scrutiny.[1] Of course, the difficulty does not concern only relativism. It also affects extreme forms of utilitarian pragmatism and instrumentalism, which identify truth with usefulness, pragmatic convenience, or practical success. Likewise, it affects exclusively descriptivist formulations of the problem of knowledge, whatever science is being used, from biology to anthropology.

Many thinkers have by now acknowledged that the limit, both of the widespread 'naturalised' or 'evolutionary' epistemologies, and of approaches exclusively based upon history, psychology or sociology, lies in the fact that they entail an unacceptable loss of the normative aspect of truth. Knowledge certainly has a biological, psychological, sociological, anthropological and historical dimension, which can and must be studied from these points of view. But, as I emphasised in Chapter II, knowledge has also a normative and justificational aspect, which cannot be given up merely because we no longer believe in the possibility of an absolute foundation, or justification, of our cognitive claims. And this ideal (normative and justificational) aspect requires the overcoming of the various conditionings, natural and cultural, which influence the cognitive process. All this is lost if we follow exclusively scientific and descriptivist epistemologies. Husserl, after having shown the sceptical implications of psychologism, went so far as to deem as *nonsensical* any claim of founding the theory of knowledge on naturalistic and factual disciplines.[2]

The weight of this kind of difficulty is considerably increased by the fact that objective truth, far from constituting an abstract ideal, is the concrete goal towards which cognitive activity tends. Cognitive activity does not limit itself to functioning within an epistemic structure: it is, in fact, guided by the intention of bringing the validity of its own presuppositions within the range of its action. An important goal that cognitive activity sets itself is the comparison between different conceptual schemes, between distant languages and cultures, between the various cultural articulations of a single civilisation, such as science, religion, myth, art, metaphysics.

---

[1] See, for example, Scheffler 1976, ch. 1, 'Objectivity Under Attack' (pp. 1-19).
[2] See Husserl 1990: 26 [33].

In other words, the effort towards objective truth and knowledge, understood as intertheoretical, interlinguistic and intercultural ideals, is demonstrated by the efforts that we constantly make to project ourselves, so to say, with the instruments offered by our conceptual scheme, beyond the linguistic, theoretical and axiological system we started with. These efforts constitute our attempt to penetrate into other systems of thought, to make them our own, through the work of translation and interpretation, and to offer, if necessary, a kind of comparative evaluation of them.

And this turns out to be all the more remarkable, if we consider that these efforts do not seem totally unsuccessful. As I pointed out in Chapter II, often the very arguments of the champions of relativism constitute a proof of our capacity of describing different languages, beliefs and cultures, with the tools offered by our conceptual scheme.

1.2. THE CONTRADICTORY CHARACTER OF RELATIVISM AS A THEORY

The second difficulty for relativism – originally identified by Plato in the *Theaetetus* – concerns its contradictory character, when considered as a proper theory asserting the relativity of truth and objectivity (as distinct from a mere relativist *attitude*). This was expressed clearly by Quine: "Truth, says the cultural relativist, is culture-bound. But if it were, then he, within his own culture, ought to see his own culture-bound truth as absolute. He cannot proclaim cultural relativism without rising above it, and he cannot rise above it without giving it up."[1] Analogously, Putnam criticises the relativist outcomes connected with forms of positivist, verificationist and naturalistic philosophies, recalling the answer that an anti-relativist such as Garfinkel is in the position to give the relativist: "I know where you're coming from, but, you know, Relativism isn't *true-for-me*".[2]

This second kind of difficulty can be easily connected with the previous kind. In the last decades, many social scientists have used their studies in order to dispute the ideal of objective truth, typical of Western scientific thought. They affirm that science as we know it is a cultural system as any other. They maintain that the linguistic, anthropological, sociological and historical study of the different forms of culture does not warrant the idea that there are apparatuses of concepts, theories and values – including those embedded in our scientific constructions – capable of transcending themselves and of being interculturally valid. The contradiction they run into issues from their claim to show in an objective way the merely intracultural validity of Western science, by relying on arguments which depend upon the acceptance of disciplines, such as history, linguistics, sociology and anthropology, which belong to the scientific patrimony of our civilisation.

---

[1] Quine 1975b: 327f.

[2] Putnam 1983d: 288. On epistemological relativism, see Siegel 1987.

It would be a mistake to consider this difficulty as the product of a negligible philosophical quibble. The human sciences too have a history, and, besides, many phases of Western culture have been strongly anti-relativist. "If we were to take social relativism literally, we would have to conclude that this view is true for a subculture of contemporary Western thinkers, period". The relativist thesis has much higher ambitions: it claims that it establishes in a "universal" way "that people's modes of cognition are a feature of their culture, not just that this is true for anthropologists" or sociologists. It becomes then "extremely difficult to see why we have a basis for adopting *this* view as universally true, while withholding such status from, for example, the view that the earth is not flat, that the stars are much farther away than many people think they are, that many diseases are caused by micro-organisms, or that people who take psychedelic drugs do not fly. It seems that those who cite anthropological data in support of the doctrine of social relativism are, in effect, reserving a special cognitive position for their own particular science, and have done so without providing any grounds for making this exception. From a social relativist perspective, they would seem to be open to a charge of extreme ethnocentrism".[1]

The same holds for relativist historicism, interpreted as the assertion that all knowledge is historically determined, and that it cannot claim a validity which goes beyond the context in which it was historically conceived. If we assign to this assertion universal validity, beyond the historical context in which it was formulated, we find ourselves in open contradiction with the assumption that there are no conceptions whose validity is not circumscribed to the situation in which they were formulated. If, on the contrary, we maintain that the validity of this assertion is itself historically relative, we automatically recognise that it is not true for the upholders of the trans-historical validity of the methods proper to modern science – as these evolved historically, in relation with different cultural practices, such as, e.g., magic, divination, or the interpretation *ex cathedra* of texts believed to be of divine origin.

By saying this, I do not intend to deny relativism any worth. On the contrary, I am convinced that our knowledge develops within variously conditioned epistemic perspectives. I only want to show the untenability of a radicalisation of the relativist thesis, which denies outright any possibility of epistemic objectivity. "If we cannot consider claims whose import goes beyond the confines of our own culture, we cannot begin to explore the role that" the linguistic, psychological and social conditionings exercise on the process of the formation of our beliefs. On the other hand, it is difficult to admit the transcultural validity of the social scientists' researches on such subjects, without acknowledging at the same time that much more developed sciences, such as physics and biology, can "provide a potential source of knowledge that transcends the culture in which these sciences first appeared".[2]

---

[1] Brown 1988: 127.
[2] Brown 1988: 130.

2. WEAK RELATIVISM AND OBJECTIVISM: TRUTH AND OBJECTIVITY AS REGULATIVE IDEALS

2.1. RELATIVISM AND METAPHYSICAL REALISM

The internal inconsistency of radical relativism is, in my view, one of the reasons why there are not many philosophers willing to proclaim themselves relativists *tout court*. Peter Machamer has adduced also other reasons to oppose epistemological anarchism, anti-rationalism, social constructivism: "I, for one, believe that avoiding these ultimate negative positions is good because it is just too depressing to be so negative. It is too humiliating to believe that one could never make a supported claim or a true statement. It is too unaesthetic to believe one cannot justifiably make any important distinctions. / Hume's claim was that the true skeptic (anti-rationalist) would have to live in a closet. Similarly no one could act on the pastiche of democratic equality that makes every individual pronouncement equal to every other, or the caricature of toleration that says that all claims deserve an equal hearing".[1]

But how can we hope to avoid radical relativism without ignoring all those aspects of cognitive activity that show its dependence on a more or less varied and structured set of epistemic conditions? The problem is philosophically more pressing when we recall that not only the results of the various human sciences but also the logical-epistemological structure of the justification of our cognitive claims require us to accept a certain dose of relativism.

It seems to me that in this case, as with other philosophical perplexities, the preliminary distinction between a 'weak' and a 'strong' version of the thesis under consideration is advantageous for the identification of the way out – or, better, of a way out.

(1) *Weak* epistemic relativism limits itself to maintaining that the cognitive process, as represented eminently by scientific activity, takes place within a provisionally accepted conceptual scheme variously conditioned by different factors. This conception is backed by the epistemological arguments discussed in Chapter II, and can be incorporated without difficulty into positive philosophy, as its relativist component.

(2) In its *strong* form, relativism consists in the further and more drastic assertion of the *non-transcendibility* of our presuppositions. Consequently, radical relativism necessarily entails the intrasystemic or intratheoretical relativity of knowledge, objectivity and truth – and not simply the contextual relativity of the justification of our specific cognitive evaluations, of our specific applications of the terms 'truth' and 'cognitive objectivity'. The strong version, that is, as well as maintaining that knowledge originates and develops within a conceptual framework, affirms also that we cannot *in principle* replace the system we started

[1] Machamer 1995: 635f.

with, that we cannot transcend our presuppositions and evaluate them in the light of alternative presuppositional frameworks relying upon a non-relativist ideal of knowledge, truth and objectivity. It is the strong version of relativism which runs into the contradiction inherent in the affirmation of the relativity of truth, and clashes with our capacity to transcend the conceptual frameworks we started with in order to describe, and adopt, others which we deem better.

Strong relativism has been contested by Quine as much as by Popper. Quine has done this by limiting himself to appending to the statement of the relative character of our alethic evaluations "the saving consideration [...] that we continue to take seriously" the more or less structured complex of our hypotheses and theories, improving it as best we can and trying to prove that it is the best we have at any given moment: "[w]ithin our own total evolving doctrine, we can judge truth as earnestly and absolutely as can be; subject to correction, but that goes without saying".[1]

For Popper, *The Myth of the Framework* is "the central bulwark of irrationalism". His "counterthesis is that it simply exaggerates a difficulty into an impossibility". The "central point" of Popper's position "is that a critical discussion and a comparison of the various frameworks is always possible". "I do admit that at any moment we are prisoners caught in the framework of our theories; our expectations; our past experiences; our language. But we are prisoners in a Pickwickian sense: if we try, we can break out of our frameworks at any time. Admittedly, we shall find ourselves again in a framework, but it will be a better and roomier one; and we can at any moment break out of it again".[2] Popper differs from Quine in that – as I said in the previous chapter – in Popper's writings the term 'better' aspires to a connection (which unfortunately did not work) with the idea of conjectural approximation to truth and objectivity, understood as metaphysically transcendent regulative ideals.[3]

Especially after the crisis of Popper's deductivist programme, many philosophers have thought that to follow the path of realist metaphysics entails emancipation from radical relativism only at the price of either accepting epistemological scepticism or compromising the cognitive value (the very claims to truth) of all those activities in which we usually engage with the intention of attaining knowledge of the object. The two alternatives have been focused on well by Hempel:

(1) lacking 'reasons', even highly conjectural, which would allow us to evaluate verisimilitude, "we never know how close we are to the truth at any point in the continuing process of formulating, testing, and adjusting our conjectures about the world";

---

[1] Quine 1970: 24f.

[2] Popper 1987: 56f.

[3] See Popper 1972b, secs. VIII and IX; 1975a: 29f, 120, 126 and 237, 317f; 1974d: 1157 (with reference to Schilpp (ed) 1974: 685; see also p. 15); 1983, p. I, ch. I, sec. 2/III.

(2) the analysis of scientific procedures shows unequivocally "that none of the considerations involved in the critical appraisal and the acceptance or rejection of empirical claims [*i.e., empirical adequacy, coherence, large scope, simplicity*, etc.] has any bearing whatever on the question whether the claims in question are true or likely to be true".

In the light of these alternatives, Hempel reached the conclusion that "the idea of science as a search for truth, for a true description of 'reality', has to give way to an epistemically relativized conception of scientific inquiry as directed towards the construction of ever-changing epistemically optimal world-views".[1]

In my view, though, both the alternatives outlined by Hempel depend upon presupposing the epistemological conception proper to metaphysical realism. Personally, I would like to stick with moderate epistemic relativism, without giving up the third great leading idea of the *esprit positif*, namely the cognitive and objective (in a non-metaphysical sense) value of scientific activity and other more or less analogous activities. Consequently, I need a conception of truth and objectivity which, without falling into realist metaphysics, will bar the radical relativist's passage from the plurality of the versions of truth and of the world, to the plurality of truths and of worlds.

2.2. IMMANENCE AND TRANSCENDENCE OF TRUTH AND OBJECTIVITY

From this point of view, I believe that Popper's attempt to build a dam against the overflow of relativism goes in the right direction, except as regards realist metaphysics. In fact, the concepts of truth and objectivity have both an *immanent* value, which manifests itself in their intrasystemic uses, and a *transcendent* value, which manifests itself in their transsystemic application.[2] If we consider only the immanence of the justification of our specific claims to truth and objectivity, we attain radical relativism, the postulation of a plurality of 'truths' and 'worlds', where each truth and each world depends exclusively upon its particular system of reference. If we react against this by considering only the transcendent aspect, and see the uniqueness of truth and objectivity as the reflection of one reality transcending the cognitive process, we end up by neglecting the epistemic relativity of our specific cognitive evaluations, making it in principle impossible to identify truth or even degrees of verisimilitude of hypotheses and theories.

I believe that we can bring about a reconciliation between relativism and objectivism by taking up part of Popper's proposal and inserting it into a perspective different from metaphysical realism. Truth and objectivity must indeed be assumed – as Popper maintained – as unitary ideals or regulative values, which transcend the particular pre-

---

[1] Hempel 1992: 48, 51.

[2] For an outline of the tension between immanence and transcendence in some areas of 20th century scientific philosophy, see Parrini, 199+a.

suppositional structures which enframe each concrete cognitive act. Their transcendence, though, is not to be understood in the metaphysical sense of an unknowable correspondence between our cognitive claims and truth in itself. It is possible to intend this transcendence differently, by placing ourselves in a *positive*, I daresay almost *operational*, perspective.

Since the common and naïve notions of truth and objectivity are permeated with realism and correspondentism, it is necessary to partly put aside what we usually mean by truth and objectivity, if we want to grasp the notion of truth and objectivity inherent in the *anti-relativist*, or *objectivist, effort* that informs our cognitive efforts (including those which constitute scientific activity). This effort informs our cognitive efforts in so far as they are *intentionally aimed* at objectivity and knowledge of the truth. It seems to me that this is, on the whole, the direction in which neo-Kantian logical idealism moved, when the validity of knowledge ceased to be measured according to a standard that transcends it, and it was acknowledged – as Cassirer says – that "[s]cience has, and can have, no higher criterion of truth than unity and completeness in the systematic construction of experience as a whole."[1]

If we start from the idea that knowledge (especially scientific knowledge) is a fact, and not a problem (in the sceptical sense), the anti-realist thesis that "we can never compare the *experience* of things with the *things themselves*, as they are assumed to be in themselves separate from all the conditions of experience" no longer leaves us with radical relativism as the only alternative. Theses proposed, although with different goals in view, within phenomenology and logical idealism, supply us with another possibility. Referring freely to these theses, it is possible to confer a non-metaphysical value on the idea of the unitary transcendence of truth and objectivity: they are interpreted as regulative ideals which guide scientific and cognitive activity towards the substitution of "a relatively narrower aspect of experience by a broader, so that the given data are thereby ordered under a new, *more general point of view*". Cognitive activity becomes then "a perpetually self-renewing process with only relative stopping-points".[2]

It is these stopping points that give content to the categories of 'objectivity' and 'truth'. This content can be specified only case by case, in the applications of the categories of 'objectivity' and 'truth' to concrete cases and particular judgments. This, though, does not interfere with the unitary transcendence of these concepts as regulative ideals which direct the research towards conceptual syntheses progressively richer in data, more articulated, and more comprehensive. In Husserl's terms, objectivity and truth become the ideal unitary correlate of a potentially infinite series of cognitive processes that intentionally aim at truth and objectivity under the guidance of the regulative ideal of the maximum possible enrichment of experiences, and of the maximum

[1] Cassirer 1953a: 187, translation modified.
[2] Cassirer 1953a: 278, italics added in the last quotation.

possible integration of the available data and of the various unifying perspectives successively adopted.[1]

The descriptions of the real world which constitute the domain of truth, are a complex of concepts, assumptions, principles, models, hypotheses, as organic and unitary as possible, though not necessarily completely free from 'local' contradictions. Cognitive efforts (especially those which constitute science) tend continually towards truth: the goal is to give conceptual form and linguistic expression to the contents of experience and to fit the resulting assertions "in a global system as harmonious, systematic and unitary as possible".

We can go on endorsing the nominal definition of truth as 'agreement', or 'correspondence', between the plane of judgements and the plane of reality, provided that we conceive the '*real world*' no longer as an entity which metaphysically transcends knowledge, but as "*the ideal limit-plane of resolution of the fragmentary manifold of experience*".[2] From this perspective, also the notions of objectivity and truth, like that of good, become purely formal ideal categories: they become 'empty buckets' which are filled at the river of history. It seems to me that this famous and rather abused metaphor of Simmel's expresses very well the tension that must constantly exist in positive philosophy between relativist and historicist instances on the one hand and objectivist and theoretical-evaluative instances on the other.

3. THE EMPIRICIST OVERCOMING OF INCOMMENSURABILITY'S THREAT TO OBJECTIVISM

3.1. EMPIRICISM AND RELATIVISM'S THREAT TO CONTINUISM

Prima facie, one might think that my suggestion meets with an epistemological obstacle in those views of history and science which have led to radical relativism.

As I pointed out in the previous chapter, even the conception of objectivity proper to Cassirer's logical idealism, no less than that of

---

[1] Husserl came to this conclusion about the 'object of knowledge' starting from an 'immanentist' perspective on the epistemological problem, according to which "[k]nowledge must be investigated in itself, in its proper essence, not in relation to a presupposed real world, and not as a fact in this world, but in pure immanence, as absolute givenness: only in this pure immanence and absolute givenness is the totality of the real world (a world which is a represented, judged, known world, and which is only by virtue of this connection the 'real world' for knowledge) 'constituted' (a word, this, which must certainly be understood in a very precise manner)" (Husserl 1987: 5). Also in Carnap's *Aufbau* the problem of 'constitution' was supposed to be ontologically 'neutral' in Husserl's sense (see Carnap's *Aufforderung zur Teilnahme an Besprechungen über den Aufbau der Wirklichkeit (Strukturtheorie der Erkenntnis)* (published in Parrini 1993: 387-9), and Carnap's letter to G. Maxwell (9.xii.1967) (published in Parrini 1993: 393); see the Foreword, above).

[2] Preti 1946: 94.

many versions of convergent realism, relies on the idea of a certain empirical cumulativism and theoretical continuism. Cassirer himself says that "[t]he one reality can only be indicated and defined as the ideal limit of the many changing theories [...] since only [by the assumption of this limit] is the continuity of experience established. No single astronomical system, the Copernican as little as the Ptolemaic, can be taken as the expression of the 'true' cosmic order, but only the whole of these systems as they unfold continuously according to a definite connection".[1]

From this point of view, Cassirer demands more of the history of science than a phenomenalist like Mach, or a conventionalist like Poincaré, who required merely a certain cumulativism on the level of empirical-observational relations. I already mentioned the issue in the first chapter, citing Frank's association of Mach's and Goethe's ideas. Therefore, I will limit myself to recalling the fact that in *La Valeur de la Science* Poincaré insisted that the change of theories cannot leave the "relations between the 'scientific facts'" unaltered, since those facts and their connections depend upon the conventional components interwoven in theoretical elaboration. But, Poincaré maintained, scientific transformations cannot affect "the relations between the crude facts", because these facts, and their connections, do not depend upon the theoretical and conventional superstructures; they constitute a sort of "universal invariant" which remains constant with respect to the variation of theories.[2]

The rejection of an absolute distinction between theoretical and observational terms threatens even this limited form of phenomenalist, or observational, cumulativism. Given the interweaving of observation and theory – not completely overlooked by Mach and Poincaré, not to speak of Duhem – the preservation of a stock of theoretically neutral empirical connections in the passage from one theory to another can no longer be guaranteed. As I said, this raises problems for all those philosophies that are more or less committed to forms of cumulativism and scientific continuism. Among these philosophies we can number not only convergent realism, but also Cassirer's logical idealism, which, unlike Mach's phenomenalism and Poincaré's conventionalism, requires a certain continuity on the theoretical level, a certain constancy of the 'objective' reference. This constancy is challenged by the conceptions which insist upon the 'revolutionary ruptures' occurring every now and then in the fabric of our presuppositions. Thinkers who have postulated the plurality of truths and worlds have done so because they are persuaded that individuals who speak very different languages live in different realities, or that 'revolutionary' changes in the disciplinary matrixes entail changes in the universe.

In my view, these considerations expose the weak point of Cassirer's 'liberalisation' of the transcendental construal of the episte-

---

[1] Cassirer 1953a: 321f.
[2] Poincaré 1946c· 338f.

mological problem. If, in order to be faithful to transcendentalism, we maintain the anti-empiricist idea (rightly criticised by Reichenbach)[1] of the "ultimate logical invariants" of experience,[2] then in logical idealism too, as in convergent metaphysical realism, we require the history of our cognitive efforts to exhibit some form of progressive unveiling of the ultimate structures of reality. It is not very important whether these structures are conceived primarily as properties of the subject, or as properties of the object, or as both these things together: it is always ultimate structures we are dealing with, and it is unclear why idealist metaphysics should be better than realist metaphysics.

It is the empiricist component of my reappraisal of the guiding ideas of positive philosophy that saves it from running the same risks as Cassirer's philosophy. In the context of empiricism and moderate relativism, the interpretation of the unitary character of truth and empirical objectivity as regulative ideals of scientific and cognitive activity loses all metaphysical significance whatsoever and allows us to look at the ironies of history with an empiricist's detachment. I can even share the spirit with which some staunch historicists utilise history in order to deride metaphysicians and (mostly second-rate) philosophers. I can do this because I have taken my precautions against the aporias of radical historicism, and I have no reason to fear that historical facts might boomerang on empiricism itself.

In the next chapter, I will consider the aspects of the question which are relevant to methodological rules and values, and therefore to scientific rationality. For the present, I will limit myself to the descriptive (so to speak) dimension of the problem, which concerns the possible empirical, linguistic and theoretical-ontological incommensurability of scientific constructions.

I think that the fact that incommensurability and radical ruptures in the historical development of science cannot be excluded a priori does not show the impracticability of my point of view in the same way as scepticism undermines metaphysical realism at its roots. The reason for this difference is that my proposal – qua proposal endowed with a strong empiricist component – does not advance 'factual' statements which claim to be transcendentally or metaphysically founded or foundable.

"Thank God I am an atheist!", I think Buñuel exclaimed faced with theological controversies. "Thank God I am an empiricist!", I think I can exclaim faced with certain obstacles which torment transcendentalists and metaphysicians of the various kinds. My idea does not require that in fact there have not been, there are not, and there cannot be deeply divergent cultures and scientific constructions, inspired by very different ways of looking at the universe. Neither does it require that there be, or there not be, a theory, which constitutes the ideal limit towards which the changing theoretical constructions appearing at the

---

[1] Reichenbach 1978d: 401.
[2] See Cassirer 1953a: 269.

horizon of history tend. In an empiricists' perspective, these occur as empirical questions, and, happily for me, as an empiricist, I am not expected to legislate a priori on them. My conception requires only that there be no valid reasons to suppose that it is impossible in principle to establish between such systems of thought a comparison, by means of which we can achieve a comparative judgement interwoven with empirical and rational motivations – where the rationality of these motivations is to be understood according to the historically conditioned, and therefore not absolute, concept of rationality which I will specify in the next chapter.

32. MEANINGS, EXTENSIONS, AND BRIDGE PRINCIPLES

I believe that some ideas that go in the desired direction bearing on the issue of theory comparison started being developed in the seventies when the discussion on incommensurability was most intense. These ideas are due especially to philosophers of science such as Scheffler, Hesse, Martin, etc.. After what I have said in Chapters II and III, it will be enough to refer to a couple of works by Michael Martin,[1] in my discussion below.

The first point that must be kept firm regarding theory comparison has already been established in the discussion of the network model: it is possible to secure the empiricist component of positive philosophy against the threat of empirical incommensurability. We cannot in principle exclude the possibility that even apparently very disparate conceptual and theoretical structures may present empirical or observational intersections which make appropriate intersystemic empirical comparisons possible.

The second point concerns the possibility of making theoretical systems, which do not have common meanings, in principle logically comparable with respect to the usual relations of logical compatibility and incompatibility, and therefore with respect to the claims of truth or of falsehood.

Let us consider, for instance, the case of two theories, T_1 and T_2, which have, respectively, the following empirical consequences:

(1) (x) (if Px then Qx);

(2) $(\exists x)$ (Px & not Qx).

Since the consequences clearly conflict, the two theories are in contradiction and, given an empirical result favourable to (2), we can provisionally discard, by *modus tollens,* T_1 as false, and accept T_2 at least as empirically adequate.

[1] Martin 1971 and 1972, with reference to the writings of Hesse and Scheffler.

Prima facie, one might think that this kind of logical comparison between theories requires the sharing of meaning, that is, that the predicative terms 'P' and 'Q' must have the same intension in the sentences of T_1 (and therefore in (1)) as in the sentences of T_2 (and therefore in (2)). If this were the case, the comparison between T_1 and T_2 would presuppose the possibility of establishing a translation manual between the different languages in which the theories and the sentences are formulated. Of course, nothing prevents this from happening. And, if the path of translation is viable, then the two theories are formulated with the same conceptual system, expressed with the same or with different phonetic or graphic forms, according to whether we do or do not apply the homophonic and homographic scheme of translation. The possibility of translation would make the assertive contents of the two systems logically comparable.

However, we must not forget that such comparability would always be relative to certain epistemic conditions of knowability. Even the manual upon which the translation is based is an instance of empirical theory. And, the same as all other theoretical systems, it too will be empirically underdetermined. Its construction will be based, as well as on experience, on certain contextually a priori assumptions, and the translations that the manual prescribes will therefore be relative to those epistemic conditions.[1]

But, more important still, it is not indispensable that there be meaning in common, in order for there to be logical comparability between statements belonging to two different conceptual and theoretical structures: intensional identity is too strong a condition to require for speaking of logical compatibility or incompatibility between two or more statements.

(1) At first,[2] the trivial point was made that logical comparison is possible even at the level of the mere extensions of the predicates. In order for there to be possible relations of logical incompatibility or equivalence between some or all of the assertions of two or more theories, it is sufficient to show the existence of referential overlaps (total or partial) between the relevant terms. This means that it would be possible to compare the assertions of two theoretical systems, formulated with non-reciprocally translatable terms. It would be sufficient to possess hypotheses capable of establishing appropriate logical relations between the extensions of their predicative terms.

I will illustrate this with the simplest case, i.e., the case in which the referential overlap is complete. The logical conflict between T_1 and T_2, asserted by their respective observational consequences (1) and (2), can persist even when the predicates 'P' and 'Q' are substituted in (1)

[1] The arguments set forth by Quine (starting from Quine 1970: 75f) to prove that the systematic indeterminacy of translation is not to be confused with the empirical under-determination of scientific theories, are irrelevant from the point of view that I am defending here. These arguments are amply analysed in Gibson 1986

[2] Martin 1971

with the terms 'P_1' and 'Q_1', having the meanings attributed to them by T_1; and in (2) with the terms 'P_2' and 'Q_2', having the meanings attributed to them by T_2. In order for this to occur, what is needed is an hypothesis which asserts the extensional equivalence between such predicates:

(3) (x) (P_1x if and only if P_2x) & (x) (Q_1x if and only if Q_2x).[1]

(2) Later on,[2] a further advance in theory comparison was made by reference to bridge principles. In order for an alethic comparison to be possible between two theories, it is sufficient to prove the existence of bridge-principles which establish appropriate relations between the entities that the terms (or some terms) of T_1 refer to, and the entities that the terms (or some terms) of T_2 refer to. This allows also the comparability of theoretical systems which differ from the point of view of their respective ontological commitments, and are therefore apparently completely incommensurable. Since this case is particularly significant from my general philosophical point of view, I will illustrate it in more detail.

Let us suppose that T_1 (in conjunction with certain auxiliary hypotheses) logically entails Pa, and that T_2 (in conjunction with certain auxiliary hypotheses) entails not-Pa. These two statements seem in conflict. Let us suppose, however, that we come to know that the two theories have completely different ontological commitments, and that the terms of T_1 refer, let's say, only to processes, and the terms of T_2 only to ordinary physical objects. Let us indicate by a_1 the a of T_1 and by a_2 the a of T_2. In this case, the term a_1 of T_1 and the term a_2 of T_2 will have neither a common intension, nor a common extension; the extension of the one will not be a proper subset of the extension of the other, nor will these extensions have a common intersection. It might therefore seem impossible to speak of a comparison and of a conflict generated by the fact that T_1 entails Pa and T_2 entails not-Pa, because in the two theories a is used in ways which are intensionally and extensionally not logically comparable.

In fact, this is not how things stand. If it is possible to find some relations between the two ontologies, the comparison is still possible. It is sufficient to discover a kind of bridge-principle that connects the ontologies of the theories; for instance:

(4) (x_1) (y_2) (if Px_1 & Rx_1y_2 then Py_2)

(the indexes indicate that the two variables range over entities belonging to different species). Given this bridge-principle, and given the statement

[1] See Martin 1971. 22-4. See also Martin 1971: 24-6 for the case of partial overlap between the extensions of P_1 and P_2 and the extensions of Q_1 and Q_2 (p. 24f for the case of proper sub-class, and p. 25 for the case of intersection).

[2] Martin 1972

Ra_1a_2, it is trivially possible to show that T_1 and T_2 clash relatively to such auxiliary hypotheses.

Martin pointed out that, faced with this, we might be tempted to object that "any two theories are comparable relative to some bridge principle". Given one or more appropriate auxiliary hypotheses, it becomes possible to establish a comparison even between "apparently incommensurable hypotheses such as 'All planetary orbits are elliptical' and 'All circumferences of pennies are circular'". "For it is a trivial logical truth that given any two consistent sentences P and Q one can find another consistent sentence R such that when R is conjoined to P and Q there is an inconsistent conjunction". But when we speak seriously of incommensurability, we are referring to problems affecting real theories, which we fail to make logically comparable and empirically testable by means of connecting hypotheses which are not the product of abstract, purely logical artifices, but have some scientific substance. "These principles must have all the various properties essential to laws: universality, truth, counterfactual force and testability. Such statements are not so easily found and in many cases it may not be possible to find them at all".[1]

3.3 INCOMMENSURABILITY: A LOGICAL AND SCIENTIFIC PROBLEM

This last specification about theory comparison does not undermine the point I want to defend. On the contrary, it constitutes its best proof. As a good empiricist and (moderate) relativist, I am not interested in guaranteeing a priori and absolutely the discovery of empirical laws capable of making very distant theoretical systems logically and empirically comparable. Much more modestly, it is enough for me to prove what I have just argued – that is, that there are no logically founded reasons to suppose that actual scientific research cannot undertake this task with the legitimate hope of being successful.

The important philosophical point is that even the comparability between different conceptual and theoretical systems – both if they look quite close as well as if they look very distant – is not absolute. On the contrary, it is relative and dependent on the background assumptions that govern our attempts to translate and to discover adequate hypotheses of connection. As Michael Devitt has noted, it is natural that "theory comparison must always involve *some* point of view about the domain in question. But this is just to say that theory comparison is theory-laden, which is true even when the most commensurate theories are being compared".[2]

Seen in these terms, the problems raised by the incommensurability thesis are no longer philosophical problems: they will appear for what they really are, namely, logical and scientific problems. The solu-

---

[1] Martin 1972: 255.
[2] Devitt 1984: 154.

tion of the comparability problems that theories raise in concrete cases will depend upon our capacity to elaborate abstract logical models of inter-theoretical relations and to develop concrete analyses and scientific hypotheses capable of guaranteeing the individuation of the qualified areas of theoretical and empirical intersection which are needed in order to carry out the proper tests. If, in the end, all attempts to build planes of comparability fail, this will only mean that, at least for the time being, one of the two theories talks, in a way which is totally incomprehensible to us, about things whose nature we do not understand. It will not mean that the theory tells us different truths about the things that we do know, nor that the things it tells us about, if they really do exist, cannot be considered as parts of the one reality which constitutes the ideal unitary correlate of our potentially infinite cognitive acts.

The trans-systemic and inter-theoretical regulative ideal of the unitary character of objectivity cannot be disputed, when this ideal is, in an empiricist spirit, disentangled from metaphysical perspectives concerning the more or less progressive and conjecturally appraisable uncovering of the ultimate structures of the known reality or of the knowing subject. Contrary to what happens in metaphysical realism, where the absence of parameters which could establish the correspondence between knowledge and reality leads to scepticism, in a positive perspective it is not necessary to specify in a preventive and general way the instruments which make the various theoretical systems logically and empirically comparable: the unitary character of truth and objectivity, posited as regulative ideals of cognitive activity, are, and must remain, empty buckets. It behoves concrete research to fill these buckets with the contents which are deemed relatively valid in individual cases. And this holds both for the construction of increasingly comprehensive hypotheses and theories, and for the elaboration of connection schemes between different theoretical systems. After all, positive philosophy does not require the indication of procedures apt to guarantee the comparability between past, present and future theories, but the proof that this comparability *is not impossible in principle*.

This may seem to some a defect, perhaps even a serious one. On the contrary, I believe it is an advantage, because the indication of specific contents does not agree with an empiricist and relativist philosophical perspective, such as the one I suggest. The specific and concrete contents must be the result of substantial analyses, motivated and directed by particular problems – problems which do not come to be in an empirical and conceptual vacuum, but in precise research contexts, variously conditioned by a manifold of different factors. It is these factors which offer the frames of reference which are essential for a satisfactory formulation and solution of the problems.

If the real problem is to discover hypotheses which are scientifically relevant, it would be really strange if there were a method which allowed us to discover them without considering the concrete contents of scientific discourse. This problem cannot be addressed in a mechanical way. Its solution will depend upon our actual scientific capacities; and our efforts might succeed or fail. But this does not rule out the fact that our attempts must have, as a regulative ideal, the goal of maximising the possibilities of reciprocal understanding between the parties involved, in the constant search for a common ground of comparison and decision.

Here we face the combined role of the analysis of scientific theories on the one hand and of logical and epistemological research on the other: it is clear that the possibilities of comparison will depend as much upon an increasingly structured and in-depth knowledge of the assertive and empirical content of the various theoretical systems, as upon the availability of abstract models of inter-theoretical connection. The problems raised by the thesis of incommensurability are not abstract philosophical questions, but matters of logical and scientific research.

From a positive point of view, characterised by empiricism and moderate relativism, to posit the unitary character of truth and objectivity as regulative ideals means to see as one's goal not the discovery of a privileged language that will be able to say everything about its referent (an idea which is justly criticised also by Lyotard),[1] but, rather, the construction of increasingly comprehensive languages, theories and schemes of connection, which may constitute a relatively neutral basis for the discussion and evaluation of alternative perspectives. If Quine's principle of the indeterminacy of translation entails that not even meanings should be considered as something absolute, completely independent of our ways of interpreting them, then also the comparison between particular perspectives and their overcoming within wider perspectives will depend upon the context of the research and will have to be construed case by case, following all the rational and empirical constraints proper to each sector of cognitive investigation. In this sense, an adequately updated positive perspective seems to be able to encompass both the idea of the unitary character of truth, conceived as a regulative ideal, and the postmodern rejection of a 'unique grand narrative' capable of transcending the limitations and historical conditionings of the cognitive process.

3.4. THE ETHICAL ASPECT: DEMOCRACY AND PLURALISM

From what I have just said it should be clear that my conception of truth and of objectivity is not 'reductionist', and has a precise ethical and practical value. Truth and objectivity are not conceived as a juxtaposition or a summation of the various points of view suggested in order to integrate the fragmentary manifold of experience. Because of its antirealist and anti-metaphysical character, my conception does not acknowledge privileged interpreters and guardians of truth and of objectivity, and consequently it entails a philosophical and cultural attitude which favours pluralism. As I will explain in the next chapter, the essen-

[1] Lyotard 1992: 83.

tial conditions for the pursuit of knowledge lie in the dialectical comparison between rival theories aimed at the discovery of new and wider common grounds for the elaboration of the cognitive synthesis. In this sense, my proposal is clearly favourable to pluralism in all social spheres, especially where scientific investigation is pursued. Preti was right when he related certain epistemological ideas to the ideals of freedom and democracy.[1]

But I firmly reject those interpretations of pluralism which – as has often been the case and often still is the case – overlook the responsibility of the upholders of particular perspectives with respect to truth and objectivity, understood as unitary regulative ideals aimed at the research of cognitive syntheses which are increasingly integrated and rich in relevant data.

4. SCIENTIFIC KNOWLEDGE, POSITIVISM, AND ANTI-METAPHYSICS

Since it is a conceptual readjustment of the notions of truth and objectivity that allows me to eschew scepticism and the contradictions of radical relativism, Russell's remark might be applied: re-definition, the same as postulation, is to demonstration what theft is to honest work. I think that the situation is somewhat more complicated; but I will be able to clarify my point of view only in the next chapter, after having said something on the nature of philosophical questions. For the present, I will go on developing my proposal, (1) specifying its relations with the positivist tradition and (2) developing its consequences concerning metaphysics and the way of intending objectivity.

4.1. THE POSITIVE SCIENCES AS PARADIGMATIC KNOWLEDGE

The main convergence between my suggestion and positivism is, of course, the idea that the positive sciences constitute a paradigmatic instance of knowledge. It is this conviction that induced me to re-define the notions of truth and objectivity, extrapolating their new characterisation from an essential trait of scientific activity (the effort to overcome all relativist cognitive particularisms), and demoting the metaphysical and correspondentist value of truth and objectivity, especially typical of the common sense view. That is, I assign definitional and normative value exclusively to the concept of objectivity and truth which, as analysis shows, actually guides the scientific and cognitive process in its effort towards objectivity and truth. I am almost tempted to say that the philosophical positivity of my proposal lies in the fact that it both keeps itself as close as possible to the data which result from the analysis of the epistemological problem, and it recognises that a defensible philosophi-

[1] See Preti 1957.

cal position can be extrapolated from these data only if some of them are 'elevated' to empty regulative ideals.

In carrying out this 'elevation', I have distanced myself drastically from the traditional idea of knowledge as the reflection of a transcendent reality, emphasising to the maximum the unifying aspect of cognitive activity. I have thus further strengthened the rationalist component of positive philosophy, a move which, as Rey pointed out,[1] characterised other moments of the historical development of positivism.

I think that I have been able to do this without sacrificing anything of the great guiding ideas of the positivist tradition. Empiricism and moderate relativism remain firmly in place, since, as I have repeatedly said, in my perspective cognitive unification is achieved by means of the construction of empirically testable hypotheses, theories and epistemic frameworks endowed with all those characters of uncertainty, provisionality and corrigibility which the empiricist tradition has always attributed to our assertions about the world. Furthermore, if the 'rationalist' notion of truth and objectivity can withstand the impact of the incommensurability thesis, this is because I show how it can be connected to an empiricist conception, which is not compelled to guarantee a priori any form of empirical, theoretical and ontological continuism and cumulativism.

All this has important consequences for many issues at the core of current epistemological discussion. I am thinking, for example, of the traditional conception of knowledge as justified true belief, of Gettier's well-known paradox, of the idea of explanation as unification, of the retrieval of some aspects of Kant's theory of judgement, of the distinction between science and non-science, of the cognitive value of metaphysical discourse. Many of the answers to these problems suggested in the recent and less recent literature, with appropriate adjustments, could be easily incorporated in my perspective. However, since for the moment my concern is only to draw the general outline of a positive philosophy, I will limit myself to specifying how my proposal remains faithful not only to relativism and empiricism, but also to the third main directive idea of the positivist tradition: the rejection of metaphysical absolutism.

4.2. METAPHYSICS DEVOID OF COGNITIVE VALUE

In the previous chapter, I have shown that metaphysics, as alleged knowledge of the absolute, or of being qua being, transcends the decision methods proper to the positive sciences, and is therefore a pseudoscientific enterprise – even if this does not mean that it is empty of meaning. Positive scientific activity does not accomplish the reflection of a transcendent object, but the unification of experience by means of the elaboration of especially built hypotheses, theories and conceptual

[1] See ch. I, sec. 2, above.

frameworks. Consequently, we cannot bestow positive scientific value on a speculative activity which claims that it gives absolute answers to questions concerning, for instance, the essential aspects of reality in itself.

But, by taking the positive sciences as the paradigmatic example of knowledge, and by modelling on them the notions of truth and objectivity, I end up by denying to metaphysical speculation cognitive, as well as scientific, value. All legitimate claims to let us know what there is and which are the properties of what exists lie exclusively with our best scientific efforts, and with analogous efforts in the various areas of practical and cultural activity. All postulations and ontological descriptions that claim to an absolute validity are in principle outside the domain of what is cognitively decidable and are therefore devoid of cognitive value.

In my perspective, this is precisely what it means to assume a positive attitude: on the one hand, we confer cognitive value only on those discursive constructions which aim towards some form of unification of experience by employing a changeable, corrigible and improvable set of epistemic conditions of knowability; on the other hand, we deny cognitive value to all discourses which claim to let us know being qua being, independently of any conceptual scheme. My conclusion is exactly the contrary of what Duhem said when he ascribed explanatory capacity to metaphysics and denied it to physics:[1] scientific investigation has a paradigmatically cognitive and explanatory value, and metaphysics, as alleged absolute explanation of the ultimate structures of reality, is devoid of cognitive value.[2]

From this perspective, traditional metaphysical disputes, such as the one between idealism and realism, lose not only scientific, but even cognitive value. It is not possible to assert or deny an absolute opposition between the subject and the object of knowledge, going beyond what holds within knowledge itself. It is empty of cognitive value to ask oneself if the dualist framework generally used in science (with certain limitations in quantum mechanics) is or is not absolutely true, reflects or does not reflect a real ontological dualism of subject and object. In order to unify the available empirical data, science, the same as common sense, adopts a dualist scheme, which recognises as existent some knowing subjects and some known objects, and this is all we can scientifically and cognitively say. Nor is there any scientific and cognitive value in the question of what in knowledge depends upon the 'real' qualities

[1] See Duhem 1962a: 10.

[2] Salmon has repeatedly set forth a conception which is opposite to Duhem's (see for example Salmon 1993: 19). In recent years, Salmon has developed the project of an integration between the causal-mechanical model of explanation and the conception of explanation as unification, distinguishing two different levels of explanation ('local' and 'global'), considered as different parts of Railton's 'ideal explanatory text' (see for example Salmon 1992: 33-9). This project seems to me admirably suited to the idea of cognitive activity as unifying activity, and to a non-metaphysical interpretation of scientific realism

of the object, and what on the 'real' properties of the subject (for example, on the subject's perceptive apparatus), if this question is pursued in an absolute sense – i.e., with the intention of going beyond what can be said on the matter with the usual scientific and cognitive procedures.

This is obviously true also for the epistemic conditions of knowability. From a positive point of view, which values highly the objection of the 'neglected alternative', the epistemic conditions of knowability are both properties of the known object and presuppositions of our specific cognitive evaluations, since this is what the analysis of the justification process shows them to be. It is empty of scientific and cognitive value to ask if they actually are, in an absolute sense, one or the other thing, or both things.

Analogously, the metaphysical interpretation of the dispute between instrumentalism and scientific realism is devoid of cognitive, as well as scientific, value. This dispute should be addressed instead as the question about whether there is any reason, of the kind usually taken into account by scientific research, to acknowledge the reality of commonly observable objects and not that of the unobservable entities posited by scientific theories. Consequently, I am unwilling to acknowledge the cognitive value of empiricist and instrumentalist positions, which accept observable objects and suspend belief in theoretical entities without setting forth scientifically relevant reasons to justify this difference in treatment. As an upholder of a positive point of view, I see no valid philosophical reasons to contest the realism at the basis of our best scientific and cognitive practices and theories; I see only reasons to contest the attribution of metaphysical value to this scientific realism.

4 3. ANTI-SCIENTISM AND DEGREES OF SCIENTIFICITY

My formulation is not a crude and naïve form of positivism. It is anti-metaphysical, but it is not *scientistic*. Putnam says that "scientism is [...] one of the most dangerous contemporary intellectual tendencies".[1] This seems to me an exaggeration: much more dangerous cultural attitudes are alive and kicking still today. Be that as it may, my proposal eschews scientism.

To be sure, my proposal characterises knowledge by taking scientific activity as a model; precisely because of this, it yields some desired anti-metaphysical implications. On the whole, however, my proposal refers to a general notion of knowledge as conceptual synthesis of experience. And it cannot be said that this characterisation entails the denial in principle of any possible cognitive value to forms of unification different from science, such as, for example, myth, religion, narrative construction, artistic expression. I believe that, especially in the case of poetic creation, it would be a serious mistake to underestimate the cognitive depth and unifying power which can issue from its capacities

---

[1] Putnam 1983f. 211.

of metaphorical construction and linguistic innovation. My proposal does not even deny the cognitive value of metaphysical discourse, provided it does not claim to give us absolute knowledge, which transcends all epistemic conditions of reference, but sets itself the task of elaborating conceptual structures which – as Quine pointed out, polemicising with Carnap on linguistic frameworks – are both instruments for the unification of experience and descriptions of the more general traits of the natural and cultural reality we live in.

Precisely because I am particularly keen on the non-scientistic aspect of my proposal, I am also concerned that my anti-scientism not be interpreted as having a wider import than I intended. When I say that in principle we cannot deny cognitive value to any intellectual or 'symbolic' construction which lays some unifying claim, I am not implying that all discourses can be put on the same level, and that there are no differences in degree between them. On the contrary: having established the exclusion of transcendent metaphysics from the domain of knowledge, and having set aside the formulation of the problem in terms of meaningfulness, I can incorporate in my proposal the 'anti-formalism' and 'gradualism' which have emerged from the age-old discussions on the verifiability principle.

There is complete agreement between my contextual relativism and the reasons which have induced a widespread scepticism about the possibility of discovering a general criterion of the scientificity, and therefore of the cognitive value, of an hypothesis. These reasons are connected especially to the acknowledgment that the judgement on the scientific and cognitive value of an assertion depends upon the specific content of the hypothesis itself and upon the background knowledge in which it is embedded. Therefore, it is not a problem which can be resolved automatically, on the basis of abstract formalistic and metahistorical requirements. As the history of science proves abundantly, questions about the degree of scientificity of an hypothesis are themselves scientific questions that must be debated and decided upon by looking at the concrete context in which they are raised. This is the only possible way to decide, for example, if an assertion leads to new empirical implications and / or to more binding systematic connections, or if it is a spurious conjecture which does not enrich in any way the already available successful theories.

In questions of this sort, it would of course be absurd to expect answers of an 'all or nothing' kind. What we can reasonably expect are only differences in degree between hypotheses which are well entrenched in experience and hypotheses or conjectures with a very weak and not clearly specifiable theoretical and empirical content. And this gradualism fits in well with the idea of knowledge as unification.

As I have said, the domain of culture is articulated in a wide variety of intellectual and representative constructions. Some of these, such as astronomy or chemistry, can be considered paradigmatically scientific and cognitive; others express in a religious, mythological or narrative form, visions of the universe and of the destiny of humanity mostly characterised by a high content of generality and imagination, and a very low empirical content; others still are more or less abstract and theoretical elaborations, whose level of scientific respectability is often considered not yet completely established (for example: psychoanalysis), or practically non-existent (for example: astrology, magic, divination, some conceptions of the paranormal). Since in all these cultural manifestations there is the tendency to conceptually unify a more or less wide and meaningful complex of empirical phenomena, we cannot establish a distinction of species, with respect to cognitive value, between these cultural manifestations and the paradigmatically scientific theories. However, we can certainly attribute to the different symbolic constructions, although only in qualitative and approximate terms, different degrees of cognitive validity, according to their different unifying power.

In the particular cases, it will be possible to speak of differences in degree, connected with the different quantity of the considered empirical data, the different number of experimental tests withstood and overcome, the different level of simplicity, elegance, logical coherence, conceptual complexity, systematic interconnection, and so on. It will also be possible to acknowledge the heuristic value of many theories, without assigning to them a particularly high degree of scientificity.

Only by taking up a similar point of view is it possible to look at the tormented problem of the status of psychoanalysis with serene impartiality, devoid of fanaticism. To deny any scientific and cognitive value to psychoanalysis would at least commit one to overlooking the varied cluster of empirical facts which it somehow strives to connect – from the trivial ones concerning the role of sexuality in individual and interpersonal psychic life, to the more specific, and at the same time more elusive ones, deducible from therapeutic sessions. On the other hand, it is sterile to barricade oneself in a dogmatic defence of the discipline at all costs, without acknowledging the methodological weaknesses of its probative apparatus, in comparison with that of other disciplinary areas with which it can legitimately associated, such as, for example, psychosomatic medicine, experimental psychology, and, in the case of parapraxes, philology and textual criticism.[1]

5. OBJECTIVITY: EMPIRICAL CONSTRAINTS ON INTERPRETATION

I will now turn to develop the consequences of my proposal regarding the problem of objectivity. The problem stems from the relativism and empiricism which characterise positive philosophy. Relativism does not allow us to connect objectivity with a transcendent reality and with a relation of correspondence between our judgements and reality.

[1] See Parrini 1988b: 176ff (on Grünbaum's evaluation of the scientific status of psychoanalysis).

Empiricism, epitomised by the rejection of synthetic a priori principles, does not allow us to reduce objectivity, as in Kant, to the universal and necessary validity of the presuppositions on which the evaluations of our cognitive claims depend.

This does not entail that in a positive perspective, which rejects both metaphysical realism and Kantian transcendentalism, it is not possible to talk about contents, that, on each particular occasion, fill up the empty ideal categories of objectivity (and truth); it only means that these contents will vary with the changing of knowledge and they will depend upon the positive empirical and rational ties, always changeable and improvable, which can be encountered in the various phases of the cognitive process.

If we take this into account, it becomes possible to remain faithful to moderate epistemic relativism without giving full credit to the thesis of the 'loss' of the world as the objective referent of our discursive constructions.[1] The thesis of the loss of the world is acceptable only if the world we are talking about is the world of metaphysical realism, but it is no longer acceptable if by means of it we intend to deny any objective value to our empirically testable statements.

The assertion of the loss of the world, radically interpreted, contrasts with the presence of precise constraints operating on the plane of empirical contents, as well as on the plane of the rational organisation of the justification process of our cognitive claims. These ties strongly restrict the import of subjectivism, conventionalism and relativism. It is the individuation of these constraints that highlights the objective value of knowledge, especially scientific knowledge, where the empirical and conceptual interdependencies are more cogent and pervasive.

A. Intra-systemic empirical constraints

The first ties we have to take into consideration are the empirical ties operating *internally* to an accepted conceptual system. Even working within a strictly relativist perspective, it seems impossible not to recognise that, once some epistemic presuppositions have been given, it is experience that determines the intra-systemic validity of our cognitive claims.

This role of the empirical material has been repeatedly stressed, in different ways, by Poincaré as much as by Duhem, while laying the foundations of a certain conventionalist relativism. After the twenties, Reichenbach especially emphasised this fact, although in the context of an epistemological conception increasingly conditioned by Schlick's idea of the purely linguistic value of coordinating definitions. Apart from this, I think we can still agree on what he said in 1928 on the relations between conventionalism and objectivity, in his discussion of the specific case of geometry and of the coordinative definition of congruence. According to Reichenbach, our statements on reality depend upon certain conventional assumptions; but, once these conventions have been

[1] See Rorty 1982.

given, it is experience which decides their validity and establishes the objective existence of certain relations between certain statements and certain other statements (for instance, between the statements that describe the behaviour of solid rods during transport and those that describe the metrical structure of physical space). We can say that "[t]he description of nature is not stripped of arbitrariness by naive absolutism, but only by recognition and formulation of the points of arbitrariness. The only path to objective knowledge leads through conscious awareness of the role that subjectivity plays in our methods of research".[1]

This component of Reichenbach's philosophy constituted a fundamental aspect (one which tends to be increasingly forgotten) of the entire neoempiricist analysis of scientific concepts and theories. One of Carnap's and Hempel's main long-standing concerns was the identification of interdependencies between the definitional and the empirical aspects of science, and the appraisal of the limitations of the degree of conventionality entailed by these interconnections. In the years which followed the end of the neopositivist movement, an analogous concern has characterised many of Grünbaum's works on the philosophy of space and time.

In more recent years, Putnam (who was one of Reichenbach's pupils) has warned against thinking that his internal realism reduces knowledge to a mere linguistic convention. Referring to Quine's criticism of analyticity, Putnam contests Reichenbach's idea that it is possible to trace a clear boundary between conventional and factual components of scientific discourse; but he is equally determined to maintain that the alternative to the idea of an absolute notion of object, independent of any scheme of reference, "is not the view that it's all *just* language". Rather, "[w]e can and should insist that some facts are there to be discovered and not legislated by us". To be sure, "[t]o talk of 'facts' without specifying the language to be used is to talk of nothing" because "the word 'fact' no more has its use fixed by the world itself than does the word 'exist' or the word 'object'". But, in internal realism, this does not indicate the total abolition of facts, but the more modest thesis, that we can talk about facts to be discovered only after the adoption of "a way of speaking, a language, a 'conceptual scheme'".[2]

As a matter of fact, one of the reasons for my long standing insistence on the appropriateness of retaining a relative and contextual value for the distinctions between a priori and a posteriori, analytic and synthetic, hinges on this aspect of the problem of objectivity. Only by means of these distinctions can we fully appreciate the import of the intratheoretical empirical constraints on our notion of objectivity. A clear grasp of the contextual value of analyticity and apriority in the process of epistemic justification, helps us focus on the synthetic a posteriori components of declarative discourse, and the role of experience in the judgements of the adequacy of intrasystemic descriptions. Only then

---

[1] Reichenbach 1958: 37.
[2] Putnam 1988: 112f, 114.

can we fully realise the truth of the saying that intelligence proposes, but experience disposes (adapting Rescher's statement, "[i]ntelligence proposes but reality disposes").[1]

From this positive point of view, I can simply disregard the absolute notion of world, giving at the same time full satisfaction to those who believe that in the positive spirit there is something philosophically 'plebeian'. Once we give up speculating on what is, or is not, absolutely real, there still remain all the familiar differences between real things and real situations, and those which are purely invented, fantasised, imagined, dreamed and (alas!) desired. In this sense, Doctor Johnson was right: the best proof of the existence of a stone is to kick it.

Poincaré could not concede to Russell that, in an *absolute* sense, the distance between Paris and London was more than one metre, because, had he done it, he would have conceded everything; it would have been no longer possible for him to uphold the thesis of the non-objectivity of congruence, and of the conventionality of geometry. Of course, Poincaré hastened to add that it would have been mad not to take into account the difference between the standard platinum iridium metre, and the distance between the two capital cities, with respect "to all circumstances of life".[2]

It is not a coincidence that the pragmatically relevant aspects are tied to the relative differences which can be determined on the basis of experience in the light of the assumptions adopted at one or more levels of cognitive activity (linguistic, theoretical, methodological, axiological). The plebeian aspect of the positive spirit is that, once reference to the absolute has been put aside, there still exist definite limits within which, inside our conceptual schemes, it is possible to speak of reality as a social construction, of the laws of nature as cultural facts, of the dissolution of texts in their interpretations, and so on.

Here, the balance that I have tried to achieve in building my ladder should be clear. *I am keen on showing the dependence of knowledge upon a conceptual framework, as much as I am keen on showing the impossibility of manipulating the facts arbitrarily by manipulating the presuppositions by means of which we talk about the facts and constitute the reference to the world's internal articulations.* In the same way as acknowledging the cognitive role of empirical data does not lead automatically to the extreme empiricist conception according to which all is experience, acknowledgment of the role of subjective presuppositions and conventional assumptions does not lead automatically to saying that everything is subjective and conventional. *Here, as in many other instances, positive philosophy finds answers of the all or nothing kind unacceptable.* In positive philosophy, we cannot answer certain questions completely a priori, without taking into consideration how the contents of knowledge have come to be concretely and historically in

---

[1] Rescher 1987: 155.

[2] Poincaré 1900: 80f

continuous symbiosis between empirical data and subjective and conventional presuppositions.

B. Inter-systemic empirical constraints: scientific change

Empirical ties do not operate only internally to a conceptual framework. They do not limit themselves to fixing the intrasystemic contents of cognitive objectivity. In the previous pages, I have repeatedly stressed the substantial role played by experience even in scientific change, although in this case the empirical ties act conjointly with the rational ones. Since we are talking about limitations which are, on the whole, commonplace and well known, a brief discussion will suffice.

(1) In the first place, as I pointed out in my discussion of the network model and of Reichenbach's critique of the synthetic a priori, experience can make contradictions, or at least tensions, emerge between the constitutive principles of a conceptual framework, making it practically impossible to retain them all. Experience opens cracks that, in the long run, result in the collapse of our theoretical buildings. And, although experience cannot influence the work of reconstruction beyond a certain extent, univocally determining which theoretical system should be adopted among a number of possible alternatives, nevertheless experience plays a primary role in the judgements of simplicity, economy, elegance and manageability which induce us to favour one conceptual framework over another. Not all logically possible conventions appear to us equally simple; in the case of assumptions with an empirical value, the degree of simplicity is not arbitrary, but depends upon the facts of experience: these are commonplaces of conventionalist and neopositivist epistemology. It is for empirical reasons, and not in virtue of a methodological *fiat,* that the movement of the hands of our clocks aligns itself in a practically acceptable way with a standard of temporal congruence, which, in its turn, fits in well with a conveniently simple formulation of our system of natural laws.

(2) Secondly, experience allows a continuous and massive accumulation of information about particular empirical facts and more or less general connections. This has played, and presumably will continue to play, a crucial role in scientific change, fostering the construction of new linguistic, theoretical and methodological frameworks, capable of more or less radically restructuring our conceptual systems. Of course, I too am convinced that the creation of new ideas is one of the main stimuli of the development of science, and that conceptual invention depends upon an autonomous inventive capacity. But I believe that it would be wrong to push this thesis to the point of denying experience any role in the creative process.

It has been legitimately disputed, both from the historiographical and from the epistemological point of view, whether Einstein was led to the formulation of special relativity by a particular experimental result capable of deciding between classical and relativistic physics. But it would be bizarre to say that the complex of the empirical acquisitions made in the physical sciences between the time of Galileo and 1905 had no weight at all in the formulation of one, or more, alternative theories!

All these empirical conditionings, however, determine knowledge only in connection with rational ties. And this brings us back to the general problem of the rationality and validity both of scientific and, as I believe, of philosophical discourse.

Chapter VII

SCEPTICISM, SCIENTIFIC RATIONALITY, PHILOSOPHY

1. RATIONAL CONSTRAINTS AND SCIENTIFIC RATIONALITY

In the previous chapter, I argued that the thesis of linguistic, theoretical, and ontological incommensurability does not exclude the possibility in principle of interpreting truth and objectivity as regulative ideals, directing our scientific and cognitive efforts towards a knowledge as unitary and intersubjectively valid as possible. Nevertheless, I argued, as empiricists, we cannot guarantee a priori that our scientific and cognitive efforts will be successful.

In my view, the most serious philosophical problem is not the difficulty of finding hypotheses which will allow the logical and empirical comparison between apparently very different conceptual schemes. As I have said, this is a scientific problem. The main philosophical difficulty arises when, after having individuated some areas of logical and empirical comparability, and after having carried out the appropriate tests, we face more or less marked interpretative divergences about the meaning to attribute to the resulting data, due to further differences in the original theoretical, methodological and axiological presuppositions. Then the thorny question arises how to resolve the conflict and which criteria should guide our choice. This is what one usually intends as the problem of scientific rationality.

In the previous pages, I have repeatedly stressed the deep contrasts which can exist between different conceptual schemes with respect to observational descriptions, linguistic uses, and theoretical and ontological presuppositions. The contrasts concerning the methodological and axiological component of scientific and cognitive activity are no less considerable. As Kuhn pointed out, they threaten the agreement on the concrete problems that a good theory should be able to solve, on what should be considered a good or a bad scientific explanation, and consequently on which are the criteria of scientificity.

The contrasts can be relative to disagreements about the system of values and rules guiding our research; but more insidious contrasts can pertain to the way of 'weighing' a given set of methodological and axiological principles in the moment of their practical application. Such disagreements highlight all the difficulties of the idea that in scientific investigation we can refer to a "neutral algorithm", a "systematic decision procedure", similar to a "logical or mathematical proof", by means of which to solve the contrasts about hypotheses and theories in a logically univocal way.

According to Kuhn, all this does not mean "that there are no good reasons for being persuaded" of the goodness of certain choices, nor that those reasons are not "ultimately decisive" for the community of researchers engaged in an investigation. "Nor does it [...] imply that the reasons for choice are different from those usually listed by philosophers of science: accuracy, simplicity, fruitfulness, and the like". Kuhn's thesis is only "that such reasons function as values and that they can thus be differently applied, individually and collectively, by men who concur in honoring them. If two men disagree, for example, about the relative fruitfulness of their theories, or if they agree about that but disagree about the relative importance of fruitfulness and, say, scope in reaching a choice, neither can be convicted of a mistake. Nor is either being unscientific".[1]

The picture is complicated further by the fact that, in big scientific changes, all the planes I have mentioned – linguistic, theoretical, ontological, methodological and axiological – tend to be inextricably intertwined. Ways of linguistic description and theoretical-ontological convictions on the nature of the object of study are interconnected with questions of method and value. In the case of Galileo, for example, *how* an object had to be studied was not independent of *what* the object was believed to be, and from the *way* in which it was described. These questions can, in their turn, interact with even more general disagreements concerning our 'image' of science and its relations with other areas of culture and society, such as metaphysics, art, religion, ideology, politics and economy.

The variable contents of the empty category of objectivity are determined, as well as by empirical constraints, also by rational ones. If we want to speak about them, we must address the problem of the rationality of cognitive choices and of the methodological and axiological principles that inspire them, and the problem of the validity of these principles. This problem coincides, in its most general form, with the traditional problem of the rational justifiability of our beliefs and of the rationality of our cognitive (in particular, of our scientific) practices.

The vastness of the subject is self-evident. In my perspective, it becomes, if possible, vaster still. As I will explain shortly, an exhaustive treatment of the matter from a positive point of view should refer to a large range of so-called 'case studies'. The concrete illustrations should involve several areas of scientific and philosophical research. Firstly, a critical evaluation is required of the arguments which have been deployed to justify the rationality of certain presuppositions (for example: the causality principle and the rule of induction). Secondly, a detailed examination should be carried out of the great doctrinal controversies of the past, and of the scientific changes that are deemed paradigmatic – such as the passage from the Ptolemaic system to the Copernican theory, or the passage from classical electromagnetism to special relativity. Finally, it is desirable that the whole proposal be tested in the light of the issues debated in the current scientific scene. An analysis of the recent experiment on parity conservation in weak interactions would be particularly helpful for my purposes. Furthermore, it would be especially

[1] Kuhn 1996b: 199f; see also 1977b, in particular p. 331f

advantageous to have a philosophical reappraisal of many historical controversies on quantum mechanics where logical, methodological, ontological and epistemological assumptions intertwine themselves, not always unequivocally, with strictly scientific hypotheses and experimental results.

Of course, here I will do nothing of the sort: the tasks that I have mentioned constitute the outline of a vast programme of analytic research which has already been embarked upon in many areas of the history and philosophy of science.

Nonetheless, I think I can still say something on the problem of rationality. It is possible to sketch and motivate quite succinctly the main points that I would like to discuss in depth in particular case-studies. This depends upon the kind of answer I intend to give. In the first place, the notion of rationality, or rational justification, that I will defend does not aspire in any way to show the possibility of proceeding without any presupposition. Less still does it seek to guarantee the unconditional validity of certain presuppositions. From the absolute point of view, Hume was right, full stop. Secondly, my main goal is precisely to show that the issue of rationality cannot be addressed in purely formal and abstract terms, with no reference to concrete contexts. In this respect, my perspective on rationality is similar to my perspective on truth and objectivity, although my answer to the former question will not be based upon a definitional readjustment.

I will limit myself to tracing the general lines of my position, deferring to other occasions the analysis of concrete cases of rational justification, relative either to discussions about general principles and their conceptual interdependencies, or to particularly significant instances of scientific change and contrast.

2. THE OPEN TEXTURE OF REASON

In the case of rationality, as in that of truth and objectivity, there is a solution which presents itself as rather obvious, but does not work. The obvious answer for knowledge was realism and the correspondence conception of truth; for rationality, the obvious answer is to trace it back to certain normative principles characterised by a specific theoretical, methodological or axiological content. In this view, reason is characterised by clearly specifiable canons which define the scientific and cognitive method of investigation and determine the rationality of choices and of decisions. This idea had a wide and long-standing philosophical and scientific following and it was shared, in various forms, by philosophers of the most disparate orientations. It has not resisted, however, a series of criticisms culminating with Feyerabend's famous attacks. The failure of the attempts so far made to solve the problem of rationality by attributing an intrinsically rational value to certain contents is by now a consolidated fact. The last failure is that of Popper's fallibilism: its methodological rules have proved themselves to be neither deductively justifiable nor descriptively adequate with respect to actual scientific practices. Moreover, according to some scholars, the disruptive impetus of methodological anarchism should be blamed on the excessive expectations imprudently raised by Popperism.

Three main objections have been made, in recent and less recent literature, against purported content-based solutions of the problem of rationality.

(1) Many philosophers have questioned the rationality of certain norms and the possibility of giving a satisfactory justification of them. It is sufficient to think of the crisis of the notions of evidence and self-evident knowledge, of the classical Humean analysis of causality and induction, of all the criticisms of Kant's attempt to found transcendentally the synthetic a priori principles of reason.

(2) The historical analysis of the supposed paradigmatic rational enterprise – science – does not support in the least the idea that certain rules impose themselves as intrinsically rational. On the contrary: it seems to suggest that there are no maxims which have been uniformly respected and about which we can say that they will never be broken or abandoned. "The history of discoveries is made by people who did not follow the rules", says Doctor Barnabas Fulton (Cary Grant) to the assistant who, in *Monkey Business* (1952), tries to dissuade him from experimenting on himself because self-experimentation is forbidden by the canons of good research. As Percy W. Bridgman said, "Scientific method consists in using your head and doing your damnedest".[1] The dadaist Feyerabend said nothing new when he referred to Lenin to maintain that there are more surprises in history than even "the best historian and the best methodologist can imagine", and that science has developed only because scientists have not let themselves be bound by allegedly immutable laws of reason.[2]

(3) Lastly, it has been pointed out that, in order to apply the supposed constitutive contents of rationality, we must presuppose and apply reason. This aspect of the question will play an important role in my argument.

In recent years, Putnam has driven home the point: the main limit of the attempt to completely naturalise and historicise reason is the fact "that the 'standards' accepted by a culture or a subculture, either explicitly or implicitly, cannot *define* what reason is, even in context, because they *presuppose* reason (reasonableness) for their interpretation. On the one hand, there is no notion of reasonableness at all *without* cultures, practices, procedures; on the other hand, the cultures, practices, procedures we inherit are not an algorithm to be slavishly followed. As Mill said, commenting on his own inductive logic, there is no rule book which will not lead to terrible results 'if supposed to be conjoined with universal idiocy'".[3] Mill's remark, incidentally, combines with Peirce's,

---

[1] Bridgman's dictum, quoted in Koertge 1980: 139.

[2] See Feyerabend 1979: 17ff.

[3] Putnam 1983e: 234

"that writers of logic-books have been, themselves, with rare exceptions, but shambling reasoners".[1]

As Reichenbach says, reason and rationality are intrinsically plastic. They cannot be codified in a series of general canons, fixed once and for all and applicable to every possible situation.[2] The open texture of reason and rationality reflects the dependence of our evaluations upon inextricably intertwined considerations of local and global factors. The choice of which maxims to apply turns upon how we consider the particularities of each case we wish to evaluate. On the other hand, the elaboration of rules depends, in its turn, both upon general considerations and upon considerations relative to possible particular applications.

The role of methodological principles is similar to that of proverbs in life: both are born from the consideration of singular cases, and both must help in the evaluation of specific situations. Exactly because of this, we end up with a proverb for each occasion and each solution; the problem is to draw suitable maxims from experience, and choose each time those which are best applicable to the concrete situations.

All this seems to entail a *capacity of judgement* which, at least in part, depends upon tacit abilities, not exhaustively codifiable in explicit and general formulations. Keynes, as well as Polanyi, had already emphasised this commonplace of the new philosophy of science. I believe it was Keynes who pointed out that the best experts in economic theories were not always the most successful businessmen, and vice versa. Analogously, those who know most about rules of conduct and scientific methodology are not always those who know best how to behave or how to find the best scientific answers, and vice versa. To compress rationality into a set of canons fixed once and for all is to prepare a Procrustean bed for what is commonly called 'intelligence'.

In the light of all this, the desire to retain some legitimacy for the notion of rationality might start to appear desperate. We might be tempted to put it aside, along with those concepts that Kant called "usurpatory [...], such as *fortune*, *fate*, which, though allowed to circulate by almost universal indulgence, are yet from time to time challenged by the question: *quid juris*". And "[t]his demand for a deduction involves us in considerable perplexity, no clear legal title, sufficient to justify their employment, being obtainable either from experience or from reason".[3]

But, despite my agreement with Schlick on the failure of Kant's transcendental deduction – and consequently on the impossibility of progressing substantially beyond Hume's point of view[4] –, I want to say a few philosophical words in favour of rationality. A passage in which

[1] Peirce 1960: 252 [3.405]).
[2] Reichenbach 1978c: 322.
[3] Kant 1985: 120 [A84 = B117].
[4] See Schlick 1974: 398ff.

Kuhn rejects the accusation of irrationalism repeatedly raised against his conception of science and scientific change will help me clarify my intentions. Kuhn repeatedly maintains that his opponents use the terms 'irrational' and 'irrationality' as "mere shibboleths, barriers to a joint enterprise whether conversation or research".[1] In his reply to Lakatos, Kuhn specifies that "[s]cientific behavior, taken as a whole, is the best example we have of rationality. Our view of what it is to be rational depends in significant ways, though of course not exclusively, on what we take to be the essential aspects of scientific behavior". This does not mean that scientists always behave in a rational way. Kuhn simply wants to say "that, if history or any other empirical discipline leads us to believe that the development of science depends essentially on behavior that we have previously thought to be irrational, then we should conclude not that science is irrational but that our notion of rationality needs adjustment here and there".[2]

We can assume two opposite attitudes towards the picture of the status of rationality which has emerged so far.

(1) We can choose an absolute point of view, like that of the metaphysical realist who asks whether the statements that we judge true are really true, and whether the entities which we posit as real are really real. In the present case, this means to ask whether our usual criteria and standards of rationality are themselves rational, and consequently whether the judgements and choices that appear to us rational on the basis of these criteria, are really such.

The inevitable conclusion is that the question of the rationality of our criteria of rationality cannot be rationally decided. In this case too, it would not be legitimate to consider our incapacity of justifying logically or empirically certain principles (for example, the principle of induction) as a proof of their non-rationality. We can only say that we are not able to decide whether these principles are rational or irrational because we do not know how to address the question of their absolute rationality. The empirical and logical unjustifiability of the rule of induction entails the non-rationality of induction only if we define the domain of the rational on the basis of the principles of empiricism and / or of the laws of deductive logic. But how are we to prove that *these* norms are intrinsically rational, and that they exhaust the domain of absolute rationality? Interpreted in an absolute sense, the problem of rationality becomes a spurious question, a question which transcends our capacity of rational decision.

And if, on the other hand, setting aside Feyerabend's caveats, we yield to the temptation of embracing the saying 'anything goes' in a prescriptive sense as the only rationally valid methodological rule, we end up in a contradictory situation, similar to that which affects radical relativism: if all is well, it is well that all is well, but it is also well that not all is well – for instance, that only what is in agreement with this or that

---

[1] Kuhn 1987b: 263; see also 1971: 139, 143f.

[2] Kuhn 1971: 144.

specific norm, is well. (Nor was this the sense in which Feyerabend set forth the principle that 'anything goes'.)[1]

(2) If we do not want to make the notion of rationality illegitimate and inapplicable, we must look for a different perspective. We must acknowledge with Kuhn that, both in practical life and in cultural and scientific life, a distinction is constantly made between rationality and irrationality. We must then examine the conditions of application and the contents of that dichotomy. Finally, we must find the way of securing a normative value for the distinction between rationality and irrationality. We can presume then that the distinction between rationality and irrationality will reveal itself to be a distinction of degree rather than a clear cut bipartition; and that some of the criteria with which we apply the distinction will turn out to be rough, very loose, analytic characterisations of it.

From the general point of view, this second perspective fits in with the conception according to which the primary and fundamental (but not necessarily exclusive) task of philosophy is not the search for absolute foundations, but the analytic clarification of the concepts we apply and of the presuppositions that, explicitly or implicitly, preside over this application. In the specific case, the task of philosophical elaboration is not to justify certain rules as absolute precepts of rational judgement: it is, rather, to clarify the notion of rationality by taking into account its actual uses and the presuppositions that discipline its application. This is just as in the case of scepticism, where the primary (although not exclusive) task I assigned to philosophy was not the neutralisation of sceptical doubt, but the clarification of the general content of the notion of knowledge and known object on the basis of the presupposition that knowledge is a fact and not a problem.

3. IMMANENCE AND TRANSCENDENCE OF REASON

3.1. NATURALISM, HISTORICISM, AND RATIONALITY AS AN EMPTY CATEGORY

In the case of rationality, as in that of truth and objectivity, I think we should start with the acknowledgment that reason is in some respects immanent in a conceptual and cultural system, whereas in other respects it transcends specific cultures and conceptual schemes. This is the philosophical lesson that can be extrapolated from many developments of analytic philosophy in the last decades.

Austin and Putnam, for example, make it quite clear that every acceptable reconstruction of the notion of rationality must be able to 'balance' two fundamental contrasting aspects: (1) it is possible to give a content to our attributions of rationality and irrationality in any sphere of practical and cultural life only by working "against the background

[1] See, for example, Feyerabend 1979: 27f and the remark in Feyerabend 1973: 146f, n. 38.

of an *inherited tradition*"; (2) it is possible to criticise and rationally modify any type of inherited cultural tradition. Austin pointed out that our traditions and linguistic uses incorporate contents which the changing of our convictions has confined to the domain of superstition, error and fantasy. Moreover, as I have already said, rationality is not definitionally reducible to its specific and context-relative contents.

The solution of the problem must therefore acknowledge that "[r]eason is [...] both immanent (not to be found outside of concrete language games and institutions) and transcendent (a regulative idea that we use to criticize the conduct of *all* activities and institutions)".[1] If we lose sight of the immanence of reason, of its relativity to specific cultural contexts, we fall into fascinating but empty "philosophical fantasies", abundantly exemplified by many speculative constructions. If, on the other hand, we lose sight of the transcendence of reason, we fall into forms of radical relativism which do not do justice to the rational revisability of traditions, as well as getting caught in the contradiction of not being rationally defensible against anti-relativism.

So, reason must not be made completely immanent and relative. Therefore, it will not be possible to stretch Kuhn's previous observations to the point of hypostatising into fixed and absolute structures the principles which embody rationality in certain 'paradigmatic' moments of the history of scientific thought. It will not be possible to identify reason with the criteria of rationality successively refined and established in the course of history, within science as well as within other complex cultural structures subject to the 'process of rationalisation', such as juridical procedures, economic structures, political institutions, etc.

If it is true that, as Mach said, history has made everything and history can change everything, we must recognise that also the opposition between rationality and irrationality is rooted in the course of events, is subject to historical change, and could even be replaced by oppositions of another kind. Nevertheless, as adequate as a 'descriptivist' and historicist conception of reason can be, it will not be possible to identify reason itself with some specific set of naturalistically and / or culturally given criteria, on pain of falling into relativism and its internal contradictions. And, if reason and rationality cannot be defined by means of these criteria, what I have said about truth and objectivity will hold for them too: reason and rationality must be treated as empty, purely formal, axiological categories, which get their content from the river of history. Only thus, I believe, is it possible to make rationality and reason both immanent and transcendent at the same time.

In this sense, my positive conception, as well as not being scientistic, is not naturalistic either. As I said in Chapter II, distancing myself from Quine, I retain the justificational aspect of knowledge, and therefore the idea that the notions of rational correctness and rightness are not reducible to biologically, psychologically, sociologically and historically determined behavioural structures. As Kant says, an adequate

---

[1] Putnam 1983e: 234, with reference to Austin 1970

conception of knowledge and rationality cannot resolve entirely the question of the *quid juris* in that of the *quid facti*.

In order to avoid the aporias of radical naturalism and historicism, I suggest we consider reason and rationality as empty axiological categories, of which we can say what Kant said of the good, namely, that in the world there is nothing good except good will. Therefore, I do not agree with Leibniz's condemnation of the Cartesian rules, even if they could really be assimilated "to the precept of I know not which chemist: take what you need, and do what you must, and you will obtain what you want".[1] As a matter of fact, if we are to avoid unacceptable forms of naturalistic or historicist relativism, we must recognise that the only defensible rule is the empty exhortation to follow reason, to be rational, in the awareness that the specific contents of rationality are closely tied to the complex of substantial information concerning both general laws and principles, and specific contextual characteristics of particular problems.

If we adopt this philosophical perspective, the question of rationality breaks up into two separate problems, treated, respectively, in sections 3.2. and 3.3., below.

(1) There is the problem of what characterises rationality in general, and therefore of what characterises the rationality of a human enterprise considered in its entirety (for example: science; the system of jurisprudence and of the practices connected with it; the ethical and behavioural code; aesthetic discourse).

(2) On the other hand, there is the question of how is it possible to rationally transcend the particular presuppositions, constituting the relative parameters of rationality in the various cultures and in the various historical phases, so as to be able to compare them in order to achieve increasingly satisfying realisations of the abstract ideal of rationality.

I admit that the generality of both these questions is such that it is almost impossible to say about them something different from the most trite precepts of common good sense. What was once said about methodology seems perfectly applicable to rationality in general: we cannot expect deep truths from it. This holds in a particularly significant way for the first question: the answer to it cannot but be the tautological explication of the empty concept of reason and rationality.

3.2. THE CRITICAL RATIONAL ATTITUDE AND REASON'S TECHNIQUES: BEYOND 'LOGICALITY'

As is clear from a neglected discussion which took place between Paul Bernays and Karl Popper a few decades ago,[2] the first problem can be traced back to two rather obvious considerations.

---

[1] Leibniz 1965. 329.

[2] Bernays 1974; Popper 1974c

(A) The first consideration is that rationality, understood as use of reason, is realised eminently in the adoption of a rational attitude in judging. That means that conflicts of opinion are dealt with by means of the ancient and venerable procedures – today commonly associated with Popper's name – of critical rational discussion.

Trivial as this may now seem, we should not forget – and historians do well to remind us – that the research of rationally motivated knowledge developed in contrast with cultural attitudes based upon mystical abandonment, fideistic acceptance of a revealed creed, acceptance of knowledge through initiation, ancient and venerable theology. To be rational means, by contrast, to judge with reflection, caution and intellectual probity, to suspend judgement in all the appropriate circumstances, to be of open mind, and to be willing to submit one's choices and evaluations to public, intersubjective discussion and testing.

The emptiness of the brief indications I have just given depends, of course, upon the fact that a genuinely rational attitude must not simply be critical, but 'rightly' or 'soundly' so. As Popper and Kuhn have variously pointed out,[1] the development of scientific knowledge would not be possible without a certain dose of faith in our ideas. I think it is clearly useless to search for general criteria with which to establish a priori when the critical spirit degenerates into frivolousness, superficiality and fickleness, and when confidence becomes dogmatic fideism.

I would like to point out the ethical dimension of the critical rational attitude. As Duhem noted, quoting Bernard, "[t]he sound experimental criticism of a hypothesis is subordinated to certain moral conditions; in order to estimate correctly the agreement of a physical theory with the facts, it is not enough to be a good mathematician and skilful experimenter; one must also be an impartial and faithful judge".[2] From this point of view, to believe in rationality means not only to assume a rational attitude, but also to commit oneself to contributing to the diffusion of the conditions of intellectual and moral life which foster the possibility of attaining the formulation of the most rational judgements we are capable of. Hence the well known connections between the development of science, the rules which govern the selection of the best experts within the scientific community, and the realisation of social and political structures apt to favour critical rational discussion.

(B) The second consideration is that rationality consists in the use of reason and its instruments, or 'techniques'.[3] Reason's main technique is, of course, conceptual elaboration and the use of arguments. Reason is realised in discursive conceptualisation and in the integration of experience according to concepts and rules of rational con-

---

[1] See in particular Popper 's remarks in 1987: 55f.

[2] Duhem 1962a: 218.

[3] The reference to 'reason's techniques' was a fundamental theme of the 'neo-Enlightenment current of Italian philosophy in the period 1945-60 (N. Abbagnano, N. Bobbio, L. Geymonat, G. Preti)

catenation such as those codified more or less exhaustively in the works on the structure of deductive and inductive arguments.

It will, also in this case, not be possible to establish a priori principles of judgment given once and for all: we must rest content with the fact that rationality can be specified only by contrast with forms of culture which exalt non-rational ways of proceeding, such as intuition, direct contact with the object, initiation, faith in spiritual authority, divine testimony, and so on.

And, once again, it will not be possible to press the opposition between rational and non-rational instruments of knowledge up to the point of denying any role in rational activities for intuition, instinct and faith. The "tacit dimension" that Polanyi speaks of is connected with the 'animal' and 'instinctual' substratum present even in scientific activity, both in its trivial aspects, tied to 'good sense', and in its more creative ones, which concern the so-called 'intuition' or 'nose' of the great scientist.[1] Duhem, for example, was very explicit in recognising a role to these faculties, in order to fill the void left by logic and experience in the selection of hypotheses and theories.[2] Upon these faculties depend both the non-strictly logical reasons that Kuhn talks about, and Bayes' 'plausibility arguments' guiding the "redistribution of probabilities over the competing theories on the occasion of the introduction of" new theories – even if we must recognise that "[t]he deployment of plausibility arguments is an art form for which there currently exists no taxonomy. And in view of the limitless variety of such arguments, it is unlikely that anything more than a superficial taxonomy can be developed".[3]

The important point is that, when we invoke capacities partly connected to instinct, intuition and good sense, we are referring to ways of judging which must be considered alogical or extralogical, but not extrarational or irrational. The idea that rationality can be identified with *logicality*, that is, with the domain of the logically provable, has been repeatedly contested in many areas of recent epistemology and philosophy of science.

Moreover, there have been various reappraisals of so-called casuistic judgements, typical of jurisprudence, clinical medicine, art criticism and historiography. Casuistic judgements are rational because, although they do not proceed by means of logical and mathematical subsumption of the particular to the universal, they can still be argued for and criticised by citing other particular cases, precedents in the application of the pertinent maxims, various kinds of analogies, and so on. In many areas of the philosophy of law it has been more or less explicitly suggested that juridical proof can be considered a paradigmatic model

---

[1] Polanyi 1958 and 1966.

[2] See Duhem 1962a, p. II, ch. VI, sec. 10, 'Good sense is the judge of hypotheses which ought to be abandoned' (pp. 216-8), and ch. VII, 'The Choice of Hypotheses' (pp. 219-70).

[3] Earman 1992. 197.

of empirical proof, and that the dialectical arguments developed in judicial controversies can be considered a particularly clear expression of the exercise of thought.

These forms of judging seem particularly relevant to the understanding of the process of scientific discovery. Reference to them can dispel, at least in part, the aura of irrationality with which Popper and the logical empiricists have surrounded it. These forms of judgement are connected to the famous Kantian dichotomy between determinant and reflective judgement, recently reappraised for similar reasons by an author quite distant from analytic epistemology, namely Lyotard.[1]

If we accept this widening of the notion of rationality, even choices and behaviour not describable by means of algorithms of deductive logic or mathematical demonstration become classifiable as rational. Hyperrationalistic projects, which aim at codifying *the whole of* reason in a system of explicitly formulated abstract and general formal criteria have failed. But this must not lead to neglecting what there is rational in the search for the optimal conditions of judgement, and in the various manifestations of our faculty of judging.

Let it be clear that these intentionally brief considerations ((A) and (B), above) are not to be taken as an exhaustive characterisation of rationality: rationality must remain an open-textured notion, whose boundaries are uncertain, not specifiable once and for all. These indications allow us to say that the scientific enterprise (understood in the widest sense, including also historiography or literary criticism) is an eminently rational form of culture. It is not so because all the episodes which make up its history depend upon reason alone, but because, on the whole, it aims – to a large extent successfully – at embodying the set of norms, ends and values which constitute the abstract idea of rational judgement. Science, as a cognitive activity, "distinguishes itself from arbitrary opinion and from faith"[2] more than other areas of culture, because of the weight it ascribes to the conceptual organisation and to the quality of the arguments of its discourse, as well as to critical and rational discussion entailing public tests based upon experiences as intersubjective as possible.[3]

---

[1] See Lyotard 1992; some of the works in which this line of thought has been pursued in the last decades are collected in Nickles 1980a and 1980b.

[2] Hutten 1976: 233.

[3] Kuhn says that "[f]ew or none of these disciplinary clusters ['fine arts', 'medicine', 'law', 'engineering', 'philosophy', and 'theology'] can be characterised by a set of necessary and sufficient conditions for membership. Instead, one recognizes a group's activity as scientific (or artistic, or medical), in part by its resemblance to other fields in the same cluster and in part by its difference from the activities belonging to other disciplinary clusters. To learn to use the term 'science', one must therefore learn also to use some other disciplinary terms like 'art', 'engineering', 'medicine', 'philosophy', and perhaps theology" The use of argument is fundamental in scientific activity, and this is connected to its rationality: "[l]ike 'mass' and 'force', or 'science' and 'art', 'rationality' and 'justification' are interdefined terms" (Kuhn 1983: 567, 569).

Thus science manages to elaborate discursive and empirical unifications which are increasingly wide, powerful, detailed and precise. To be sure, the past and present practice of scientific communities does not always conform to the axiological ideals of rationality; but this fact cannot obscure the inspiring principles of the scientific spirit, nor the momentous quantity and quality of its results.

With this, I do not intend to automatically uphold the superiority of scientific culture and rational attitude on differently oriented forms of culture and attitudes. Personally, I am inclined to agree with Mephistopheles, when he says to himself that, in order to go to ruin, it is not necessary to sell one's soul to the devil, as Faust is doing: it is sufficient to despise "reason and science, supreme powers of mankind". However, nothing of what I have said so far claims to justify, not even relatively, the superiority of the idea that in life it is better to rely on reason and on its scientific realisations, than to be irrational and confide in instinct, faith or mystical abandonment.

3.3 SCIENCE AS A SELF-CORRECTING PROCEDURE: TRANSCENDING CONCEPTUAL SCHEMES

This last question belongs to the second aspect of the problem of rationality, which is by far the thorniest. The issue is the possibility and the ways of transcending our own standards of rational evaluation, and consequently also the comparability of our standards with other more or less different cultural perspectives with a view to choosing the best.

Despite the breadth and the difficulty of the subject, I will be quite concise, since many recent studies adopt an openly anti-foundationalist perspective, similar to the one I want to uphold, and which is incorporated in the network model introduced in Chapter III. The ideal epigraph of these studies is Neurath's metaphor of the sailors – an epigraph, in my view, implicitly inscribed in the neoempiricist criticism of Kant's synthetic a priori. This striking image conveys the idea of a relative, secularly human and this-worldly justifiability in a way that could be improved only with difficulty.

Many of these works start from the conviction that all cognitive and evaluative activity (in particular, scientific activity), is characterised by a self-correcting procedure, which is founded upon itself: this means that knowledge relies upon certain provisionally accepted beliefs to evaluate rationally and empirically other beliefs, in a virtually endless process. In this self-governing cognitive and justificational process which has no props external to the process itself, the most general and the most particular cases are inextricably intertwined and interconnected.

Let us take the classical example of the principle of induction. Let us grant for the sake of argument that the failure of the many attempts made so far of justifying induction on the basis of deductive logic and experience shows the impossibility of such an enterprise.

From a positive, anti-foundationalist perspective, the surrender of an absolute justification of induction does not imply that the great amount of work which has been done and which is still being done on the subject does not have philosophical, as well as scientific, relevance. It is sufficient to think of the works of Keynes, De Finetti, Carnap, Reichenbach, Salmon, Shimony, Hintikka, Burks, Levi, etc., in order to realise how much this effort has contributed to the kind of relative justification the value of which I want to stress. These analyses, despite being inspired by the most disparate philosophical principles, have contributed to mapping the web of complex interrelations that connect the principle of induction with the set of more or less general concepts and principles that constitute the fabric of our common and scientific beliefs.

As I noted in Chapter II, the idea of inductive confirmation is connected with that of probability. According to our interpretation of the latter notion (subjectivist, frequentist, logical, statistical, propensity, etc.), induction will turn out to be linked to quite remote ideas, concerning not only the uniformity of nature, the existence of causal links, and determinism, but also the conditions of prediction and the assumption of a limit of relative frequencies, our intuitions about the plausibility of hypotheses, certain structural characteristics of reality (such as the existence of a limited variety of species, the substantial knowledge by means of which we distinguish in experimentation between relevant and non-relevant variables, and so on).

Moreover, if we take into account the distinction between local and global induction, and we consider, as well as the connections between inductive confirmation and other general notions, also the specific contextual conditions in which inductive reasoning is used, the picture of the interdependencies becomes increasingly articulated and complex. This undermines the claim of attaining absolute foundations, while reinforcing the tendency towards a relativised conception of the justification of our empirical cognitive claims. The justification of knowledge is interpreted as a self-governing and self-correcting process, in which "we use the knowledge that we think we have already in order to gauge how good a title new evidence gives us to claim further knowledge", and we attribute an empirical and corrigible value to inductive evaluations regarding "how well the experimental evidence supports a hypothesis".[1]

In quantum mechanics, the principle of induction itself is connected with issues such as determinism and causality which, in their turn, show themselves to be tied to more or less specific scientific statements, such as the principle of exclusion, the principle of locality, the admission of hidden variables, the use of non-standard logics, and the realist assumption itself.

In biology too there are analogous interconnections: some versions of evolutionary epistemology allow us to 'justify' our instinctive propensity towards inductive behaviour by considering it as the result of natural selection, as such presumably enjoying a high survival value with

---

[1] Cohen 1988: 95; see also 1989.

respect to the natural environment we live in. This justification, however, is trivially circular, since evolutionary theory is accepted precisely because it is inductively confirmed. The evolutionary 'justification' of induction cannot claim an absolute value; but it cannot for this reason be considered less instructive from the point of view of the effort towards the most unitary organisation possible of our beliefs and convictions.

In the manner sketched above, it becomes possible to give an increasingly systematic and interrelated structure to our evaluations, thereby increasing the rationality of the system we operate in. At the same time, it becomes possible to lay down the premises for transcending the specific rationality criteria adopted on each particular occasion for the comparison of diverging perspectives and for the rational solution of conflicts based upon the identification of common presuppositions.

The systematic enrichment and the articulation of the presuppositional systems can lead us to better specify the various criteria of rationality, to identify increasingly wide networks of interdependencies, and increasingly general and powerful rules. By means of these interrelations, we will be able to verify, within the specific contexts, whether among divergent attitudes and conceptions there are any relatively non-problematic points of convergence to use as a basis for the rational resolution of contrasts.

In a critical rational comparison thus conceived, an important role will be set aside for 'elenctic demonstrations', such as Aristotle's of the principle of non-contradiction, as well as for the individuation of performative contradictions and self-contradictions, objections *ad hominem,*[1] and for the institution of more or less analogical connections between the various fields of culture (for instance, in Italy, ideological and political considerations were used to defend the principle of empiricism).

In the case of divergences with an empirical import, the rational constraints will interact with the empirical constraints I mentioned at the end of the previous chapter. The discussion will extend in search of areas of observational interaction, availing itself of all that the empirical data will allow us to say on the internal coherence of a presuppositional system and on the best way of systematically organising the experience accumulated. We could say, freely paraphrasing Heidegger, that it is only in the 'contest' that we can hope to bring to light, as well as the presuppositions of the 'contesting' also the possible areas of agreement which can offer the premises for the rational solution of the contrast.

---

[1] The discussion on the problem of the 'ultimate foundation', as reproposed by K.O. Apel, may have a certain relevance from these points of view. See also the argument on inductive scepticism in Earman and Salmon 1992: 94

3.4. BEYOND FOUNDATIONALISM AND ANARCHISTIC NIHILISM: THE BALANCE BETWEEN DESCRIPTIVISM AND NORMATIVISM

Of course, we cannot guarantee a priori that it will always be possible to find a common ground of discussion and comparison. The important point is that it is not possible to prove the contrary. In order for relativism (the statement that our rational evaluations depend upon provisionally accepted norms of rationality) not to transform itself into the radical and contradictory thesis of the non-transcendable relativity of reason, all that is needed is to show that the search for convergence between different cultures, traditions and orientations is not in principle impracticable or doomed to failure.

It has sometimes been said that the nihilistic aspects of Dostoevsky's thought and, in certain respects, of Nietzsche's, have as their presupposition a background of negated religion, the loss of the absolute character of the divine. There really does seem to be an analogy between what could be upheld in ethics on the basis of saying that if God is dead then everything is permitted, and what could be maintained in methodology and epistemology on the basis of the contention that, if there are no rules and principles justifiable in an absolute way, anything goes.

Both nihilism and anarchism, however, turn out to be unjustified conclusions: once all values and principles have been put on the same plane from the point of view of absolute foundation, there still remain all the relative differences that human beings are capable of establishing between them. These differences concern human beings' personal propensities and the necessity of harmonising and critically comparing one's values and principles with the orientations of others. Goodman has rightly pointed out that what saves human knowledge from sheer nihilism is its being the product of an unceasing dialectical comparison between more or less radically contrasting positions.[1]

A poet had already sensed this: if there is no God on earth, courage: we will be, as best we can, gods. If there is no 'God's eye' with which to look at our theoretical, methodological and axiological assumptions, and with which to evaluate them from the point of view of their absolute correctness, then the parameters which we use in the concrete cognitive process will be the relative references for evaluation.

Thus my idea of rationality is neither wholly descriptive nor wholly normative. I do not conceive rationality in an entirely descriptive way because, against radical naturalism and historicism, I keep firm the justificational dimension of cognitive activity, and I make the justification process depend upon theoretical, methodological and axiological assumptions. On the other hand, I do not intend rationality in an entirely normative way either, because I consider those assumptions as merely relative stopping points – not foundable in an absolute sense – of the concrete epistemic processes of justification, and of the tension towards

[1] Goodman and Elgin 1988, see also Goodman 1978 and Putnam 1983c.

the identification of a possible common ground between divergent perspectives.

This balance between descriptivism and normativism allows us to overcome the traditional alternative between viewing history as justificatory or condemnatory. History does not necessarily justify, because the fact that things went in a certain way does not justify that it was right (in a relative sense) that they went as they did. A thorough historical analysis might show that the rules that were allegedly being followed were not applied well, or that there was the concrete possibility of elaborating alternative points of view and / or of using different evaluative principles which, if applied, would have led to different and more satisfactory solutions. On the other hand, history does not even necessarily condemn, because we do not assume that it is characterised by a progress somehow inscribed in the order of things, which will lead us to discover evaluative criteria certainly superior to those previously used.

Accordingly, my idea of rationality views as anachronistic and anti-historical the application to past cases of criteria of evaluation and choice which have prevailed subsequently, if those criteria could not have been formulated and appreciated in the concrete theoretical, methodological and axiological horizon of the past context under consideration. I am not saying that all evaluations of choices made in the past are automatically absurd. The point is that evaluations of the past should be made on the basis of historically adequate parameters. I do not believe that we can justify a priori any form of historical pessimism or optimism: the past is what humanity has been able to do; the future is what humanity will be able to do; there is no a priori guarantee of progressive improvement or of regressive deterioration.

4. ANTI-FOUNDATIONALISM AND THE STATUS OF PHILOSOPHY

In my perspective, this conception of truth, objective validity and rationality can be applied not only to scientific, but also to philosophical discourse. Between science and philosophy there are certainly many differences, but I think that in principle philosophy can aspire to the same type of objective validity towards which scientific activity tends, despite the fact that it is not difficult to foresee that, in many ways, philosophy will not attain the same degree of incontrovertibility and consensus. Thus, my proposal aspires to achieve a rationally valid result, in the sense of rationality and validity that I have specified. But, in order to clarify this aspect, I must first keep the promise I made in the previous chapter: I must say something about the nature of the answer that I have given to the problems of truth, objectivity and rationality.

There exist many kinds of philosophical answers to philosophical perplexities. In the case of the problem of knowledge, there are some speculatively very deep answers which, apart from their overall validity, have brought to light previously unsuspected aspects of epistemological problems (it is sufficient to think of Kant's transcendentalism and Hegel's idealism). Other solutions have been suggested which are apparently simple and linear, and hinge on *prima facie* very plausible and powerful principles. I am thinking especially of the attempts to eliminate the problem, qualifying it either as devoid of sense, on the basis of the verifiability principle, or as linguistically incorrect, on the basis of reducing meaning to the use of words. There exist also answers which conceal emptiness of thought and conceptual confusion behind feigned speculative depth.

Of course, I hope my suggestion is not of this last kind; but it is certainly not of the first nor of the second type. My main aim has been to keep myself as close as possible to the results of the analytic decomposition of the epistemological problem, even if at a certain point I had to acknowledge that I could not remain faithful to that assumption to the end. The analysis of the epistemological problem issues in an internally aporetic situation and, if we do not want to stop at that, we must proceed to a kind of synthetic recomposition that yields notions free from internal and external difficulties. Thus, in setting forth my proposal, I performed what I would call a task of 'conceptual engineering', or, in Herbart's words, of *Bearbeitung der Begriffen* (the 'reworking of concepts').

4.1. LOGICAL PARADOXES AND EPISTEMOLOGICAL SCEPTICISM: AN ANALOGY

I will elucidate my position by means of a comparison with class theory, since I believe that epistemological problems are in some way analogous to those of logic. I am thinking of the paradoxes which emerged in the naïve theory of sets (certain distinctions between classes and sets are here irrelevant), in particular of Russell's (and Zermelo's) antinomy of the classes that at the same time are and are not members of themselves. It is with reference to this contradiction that some logicians and philosophers talked about a 'hitch' in the commonly used apparatus of logical principles, affecting in particular the so-called principle of comprehension, considered as an apparently 'obvious' way of positing abstract objects, such as classes or attributes, in correspondence with each predicate. This perspective is particularly clear in some observations of Russell and Quine which, beyond their strictly logical scope and validity, have, in my view, a considerable general philosophical import.

Russell explained that the main motivations which guided him in proposing a logical system incorporating the simple and the ramified theory of types was the need to find a 'way out' from the contradictions of set theory: "[w]hile I was looking for a solution, it seemed to me that there were three requisites if the solution was to be wholly satisfying. The first of these, which was absolutely imperative, was that the contradictions should disappear. The second, which was highly desirable, though not logically compulsive, was that the solution should leave intact as much of mathematics as possible. The third, which is difficult to state precisely, was that the solution should, on reflection, appeal to what may be called 'logical common sense' – i.e. that it should seem, in the end, just what one ought to have expected all along".

Russell goes on to remark immediately after that the last requirement is not respected in the alternative solutions elaborated later by Quine on the basis of the idea of 'stratification'. For Russell, Quine's logical systems are admirable "on account of their skill", but they are not wholly satisfying from the point of view of ordinary logical intuition. In contrast with the 'naturalness' of the intuitive idea of the hierarchy of types, Quine's systems look like constructions based upon principles "created *ad hoc* and not to be such as even the cleverest logician would have thought of if he had not known of the contradictions".[1]

Russell's remarks show that he did not realise that his objection had somehow been dealt with in advance by the way Quine motivated his choices. Quine recognised the artificiality of his axioms of sets in contrast with the naturalness of the principle at the basis of the theory of logical types, but at the same time he pointed out the well known inconveniences of Russell's solution. According to Quine, the initial naturalness of the hierarchisation of the universe of objects is paid for quite dearly, since it leads to "unnatural and inconvenient consequences", such as the "reduplication", in correspondence with each type, of the universal class, the empty class, numbers, arithmetic, and so on. Only by recurring to manifestly artificial stratagems is it possible to avoid the oddities. The naturalness of the initial principles of Russell's system is paid for by the great unnaturalness of the "intuitively repugnant" "cleavages and reduplications" present in its consequences. In Quine's systems, the contrary is the case: the starting points are artificial, and the consequences are natural. Consequently, Quine's *overall* result seems preferable to that yielded by the theory of types.[2]

On the basis of this comparison, Quine concludes that, apart from the relative merits of the various proposals, the contradictions of set theory, and the work done to eliminate them, show "clearly enough that the previous naïve logic needed reforming", and that strikingly "none of these proposals, type theory included, has an intuitive foundation. None has the backing of common sense. Common sense is bankrupt, for it wound up in contradiction. Deprived of his tradition, the logician has had to resort to mythmaking. That myth will be best that engenders a form of logic most convenient for mathematics and the sciences; and perhaps, it will become the common sense of another generation".[3] (The reference to myth should not go unnoticed: as is known, Plato often resorts to the narration of a myth, in order to solve the difficulties highlighted by Socratic questioning.)

---

1 Russell 1975: 79f.
2 Quine 1964b. 91f.
3 Quine 1995: 26f

Gödel was to make a similar remark some years later: "[b]y analyzing the paradoxes to which Cantor's set theory had led, [Russell] freed them from all mathematical technicalities, thus bringing to light the amazing fact that our logical intuitions (i.e., intuitions concerning such notions as: truth, concept, being, class, etc.) are self-contradictory".[1]

I believe that analogous considerations can be made about the epistemological problem. The sceptical stance shows the internal difficulties of the naïve conception of knowledge which starts from the preliminary ontological duplication between subject and object of knowledge. On the other hand, the contradiction that relativism gets entangled in shows the unsustainability of the equally naïve suggestion of overcoming scepticism by annihilating *sic et simpliciter* the dualism, and by making object and truth internal to the various cognitive processes, so as to take all transcendence, normativity and unity away from them.

Therefore, it looks as if a satisfactory solution on the one hand must uphold the unitary transcendence of truth and objectivity in order to eschew the unsustainable aspect of relativism. On the other hand, it must not interpret truth and objectivity in an absolute way, so as to avoid metaphysical realism and correspondentism. A metaphysically conceived object would be in principle unreachable by the cognitive process. Nor is it what we actually deal with when we commonly talk of objective reference: metaphysical realism clashes with the valid aspect of relativism, that is, the dependence of the cognitive process upon an epistemic framework.

The analogy between the problems of class theory and epistemology can be extended to answer these problems: in both cases, a solution, as well as eliminating the conceptual aporias, should try to keep itself as close as possible to the non-problematic aspects of common sense which we have no specific reasons to doubt.

4.2. SCIENCE, PHILOSOPHICAL ANALYSIS, AND 'CONCEPTUAL ENGINEERING'

From the analogy between the theory of classes and epistemology, it should be clear why I do not believe that there is a difference in principle between scientific and philosophical discourse. There will certainly be a difference in the degree of generality and controvertibility, but both aim at formulating a conception of natural and cultural experience as integrated and as unifying as possible.

This affinity induces me to reject the traditional idea that philosophy should search for a certainty and a definitiveness superior to those we are willing to accord to the acquisitions of science. Therefore, I agree with the contemporary tendency to deny to so-called philosophical explanations all claims to demonstrative cogency. The extensive work of analysis carried out in the last decades has yielded conclusions

[1] Gödel 1990. 124.

on the nature of philosophical problems and solutions which are at odds with the traditional persuasion that the task of philosophy is to produce demonstratively founded conclusions that impose themselves as the only valid and acceptable ones.

Rather paradoxically, philosophy has been both the domain of endless controversies and the discipline which has retained longest the illusion that it can say the final word on humanity's most fundamental general problems – if only to proclaim their lack of sense on the basis of general formulae held to be undiscussible, such as the verifiablity principle. In recent years, partly because of the difficulties encountered by analytic approaches, a different way of looking at philosophy has asserted itself. This point of view is now upheld with all the methodological awareness and cogency that analytic philosophy has made us capable of. This perspective cannot be considered entirely new, and, in order to illustrate it, I will start from a philosopher of the past.

Herbart maintained that philosophy can be considered neither Kantian 'rational knowledge by concepts',[1] nor Fichtian *Wissenschaftslehre*. Rather, philosophy should be seen as a kind of 'conceptual elaboration'. The task of philosophy should not consist exclusively in the "logical" work "of separating and ordering concepts", that is, if we want, of *analysing* concepts in order to make them somehow clear and distinct: the task of philosophy comprises also a kind of systematic integration of concepts, attained by "reworking" them.

Herbart expounded these ideas while defending a form of realism, and referring to the internal difficulties of the common concepts of substance, causality, change, I (it looks like the index of a recent volume on "philosophical explanations"!). Apart from this, I am interested in defending Herbart's general idea that philosophy originates from the observation of aporias and difficulties inherent to certain notions and their reciprocal connections as they occur "in experience, and in particular in the experience of common consciousness". The task of philosophy is, in the first place, the clarification of concepts, even in their problematic aspects, and, subsequently, their elaboration and integration in forms as coherent and systematic as possible.[2] This, after all, is the idea underlying Carnap's well-known conception of 'explication', also connected to the project of a rational reconstruction of the basic notions of epistemology (inductive confirmation, meaningfulness, probability, analyticity, etc.).

The fundamental purpose of this book is a rational reconstruction of the key concepts of epistemology: truth, objectivity, rationality. Following Herbart's idea of philosophy, the analysis of apparently well established concepts and facts (for example, a certain logical structure of empirical tests) is followed by a rational synthesis, whose goal is the formulation of a conceptual network as coherent and comprehensive as

---

[1] Kant 1985 577 [A713 = B741]

[2] See Herbart 1964, §§ 1, 4, 5, 6, 7, General Note to ch. II of p I, §§ 117, 119.

possible. This rational reworking, as in Herbart's intentions, should not be removed more than is strictly necessary from the common view of things, which is an obvious and indispensable framework for all our pragmatic choices and evaluations.

I am fully aware of the partly synthetic character of my proposal, as well as of the impossibility of justifying or grounding it in an absolute way. And this generates a convergence between my way of looking at philosophy and some recent formulations in which philosophical research is no longer assimilated with the construction of a sort of demonstrative "tower", with the attempt "to deduce" a "total view from a few basic principles" founded on intuition (such views were already criticised by Neurath). Philosophical research is compared instead to the building of a sort of Parthenon, in which "[f]irst we emplace our separate philosophical insights, column by column; afterwards, we unite and unify them under an overarching roof of general principles or themes". All this implies abandoning the "coercive" claims of philosophical speculation. We renounce the idea of philosophy as a field of investigation in which, by means of suitable demonstrative arguments, we try to found, and then to 'impose', certain conclusions. We give up completely the traditional claim that successful philosophical arguments are strong arguments, in the sense of having the capacity to "*force* [...] someone to a belief".[1]

Personally, I still believe that there are various arguments, connected to the properly analytic aspect of philosophical discourse, which have a considerable 'coercive' strength – a strength similar to that of the demonstrative sciences. But, as Quine clearly saw, in philosophy there are, as well as the analysis of concepts and philosophically relevant 'facts', also positive proposals made up by conceptual integrations analogous to those which characterise the more abstract and theoretical parts of the scientific disciplines.

I too think that the idea of a conceptual organisation as unitary and coherent as possible cannot be justified by appealing to irrefutable prime principles. This idea can only aspire to a persuasiveness which can be assessed on the basis of its *holistic* characters, as possibly the best global reconstruction of the manifold in which the complex of our natural and cultural experiences is analytically decomposable. In this sense, no suggestion of rational reconstruction, including the one I defended, can aspire to being the only possible one and to being valid in a coercive way, even if it can acquire considerable strength on the basis of the arguments that it can muster against rival systematisations. Accordingly, my idea of philosophy is not 'coercive', but it attributes great importance, as well as to analytic arguments and decompositions, also to its claim to an objective validity to be tested on the basis of comparisons with other general conceptions.

For a philosophy so conceived, there is no particular problem of the beginning. There must be place in it for the consideration both of

[1] Nozick 1981 3f.

aporias and problems inherent in our conceptual system, and of those complex historical facts which interact with our structure of thought and modify it. In the case of the problem of knowledge, for example, it would be absurd to limit ourselves to an analysis of the common concept of objectivity, without taking into account the momentous development of an undoubtedly cognitively relevant cultural phenomenon, such as modern science. As any other kind of investigation, philosophical research deals with problems which exist within a 'web' of concepts, beliefs, and natural and cultural phenomena – a web that often presents aspects which are not easily harmonisable.

Furthermore, the fact that the great cultural formations, such as religion, science, ideology, are subject to unceasing historical transformation can explain why philosophical questions, in some respects always identical, are continually enriched by new components that challenge past organisations and require more up-to-date ones (let us think of the bioethical problems raised by the conquests of biology, or of the developments of the physical and biological sciences which, according to some, repropose the problem of the social control of scientific research!).

All this does not fit in well with a conception of philosophical problems as 'linguistic cramps', born of a 'holiday' of language connected with erroneous analogies, grammatical distortions, semantical category mistakes, and which would dissolve as soon as we are able to restore the correct linguistic usage. As well as with questions of this kind, philosophy deals also with genuine and real problems, born of objective conceptual and cultural tensions, and requiring answers on the same level.

4.3. THE PHILOSOPHICAL SIGNIFICANCE OF SCEPTICISM

This is the case for the problem of knowledge. The problem of knowledge does not have an exclusively linguistic basis, and it cannot be solved in purely linguistic terms. It is born instead from the analysis of the process of knowing and from the aporias inherent in our common notion of object, which presents itself in certain ways as transcendent, and in other ways as immanent to cognitive activity.

The question of knowledge arises already at the level of everyday experience, in relation to scepticism. It undergoes a further enrichment and cultural deepening in correspondence with the development of modern science, which adds to the problem of sceptical doubt that of harmonising the scientific accounts of the world with the image suggested by common sense. It is undeniable that many scientific constructions give us descriptions of the universe and speak of entities which, at least *prima facie*, are in sharp contrast with the common sense and everyday view of reality (Eddington's famous two tables!).

Of course, we can always avoid taking such problems seriously. But this does not make the objective depth of these problems disappear.

I will try to explain my position reconsidering the theme of scepticism and the analogy between epistemology and the theory of classes.

In recent years, someone suggested that a mathematician who was willing to go on doing his job unperturbed, could face the set-theoretical paradoxes in a wholly legitimate way by putting them aside with a smile. This is what scientists commonly do in the case of scepticism, and I do not think that science has been the worse for it. But then, why not assume the same attitude with respect to the same problem in philosophy? No sooner said than done. Recently, a philosopher strongly convinced of the validity of metaphysical realism has maintained that, in an adequate realist reconstruction of the problem of knowledge, we can put aside the sceptical stance as eccentric or irrelevant.[1] The role of scepticism in my proposal is in certain respects wholly analogous, and in other respects completely the opposite.

The analogy consists in the fact that I assign no *philosophical*, not only no pragmatic, weight to the sceptical stance, in so far as scepticism should lead us to doubt the soundness of our best established knowledge, and to set forth adequate counterarguments. In this sense, my perspective is radically anti-Cartesian. Generalised doubt is something to set aside with a shrug of the shoulders: there is no better way to overcome doubt than to resort to the efforts we usually make to silence our doubts. In a positive perspective, the common sense reply to sceptical doubt is right not only on the practical plane, but on the theoretical one as well.

The right attitude towards philosophical scepticism is the properly scientific attitude. Modern science does not overcome this kind of doubt and certify its own legitimacy by going in search of indubitable certainties: it overcomes scepticism by setting it aside, and implicitly showing its irrelevance by means of an adequate limitation of the range of its own statements.

The generalised sceptical stance depends upon the 'generic possibility' of making mistakes or of being deceived, which can be invoked in any circumstance. This possibility "relies on a general, always-available doubt, independent of any specific mode of instantiation or mechanism of operation. [...] Real science has evolved a strategy to deal with" this kind of generalised sceptical doubt: "(1) [a]dmit the possibility; (2) see whether in context it can be articulated (i.e., made specific, concrete, and testable); (3) if not, then bracket it off as a *mere* possibility and proceed with your work, returning periodically to (2)".[2]

Distancing ourselves from the Cartesian tradition, we must clearly distinguish – with Shapere – between the possibility of doubt and the reasons for doubting, understanding that "the possibility of doubt arising is not itself a reason to doubt of any particular proposition; more exactly, it is no reason to abstain from using the best beliefs we have –

---

[1] Devitt 1984: 52, 227.

[2] Fine and Forbes 1986: 238.

those which have been found to be successful and free from *specific* doubt – to build on".[1]

If we interpret sceptical doubt as an effective doubt about the existence, let us say, of "a kitchen table... when you are not there", I agree fully with those who maintain that, in its traditional specific applications (perceptual illusions, distinction between dream and reality, etc.), scepticism can be considered wholly irrelevant and set aside with a smile.

This is where the analogy ends, because, in my view, this thesis must not be taken to the point of saying that the possibility of doubt has no philosophical relevance. This would be like saying that, since set theoretical paradoxes are not important for the specifically mathematical work, they have no relevance even for those who try to understand and develop the theory of classes. The possibility of doubt is relevant for the comprehension of the structure of the cognitive process. It makes us understand that our cognitive activities can ignore sceptical doubt precisely because in fact they operate, in each particular instance, by means of the application of intrasystemic, not transcendent, criteria for the justification of truth and objectivity. Therefore, while my formulation affirms the irrelevance of scepticism to discredit our best confirmed specific beliefs, at the same time it takes seriously the possibility of sceptical doubt, with regard to what it shows about the nature of knowledge and of the known object.

If authentic culture must contain some dose of self-irony, and if philosophy is part of culture, then philosophy must have an auto-ironic component. It seems to me that a positive philosophy manages to blend quite well the seriousness of philosophical commitment with a certain ironic detachment from purely speculative problems.

5. CONCLUSION

As I have said, in evaluating my proposal one should consider its claim to being a reconstruction of the phenomenon of knowledge as unitary and integrated as possible, and therefore objectively and rationally valid. Therefore, I believe that the objections variously raised against the positivisms of the past cannot be raised against it.

As especially Putnam has recently insisted, one of the most serious limits of verificationist, positivist and naturalist philosophies is a defence of conceptions of rationality and / or meaningfulness which are "self-referentially inconsistent", that is, "*so narrow as to exclude the very activity of producing*" those conceptions.[2] It is sufficient to think of the famous, radical objection, according to which the verifiablity principle is not verifiable, and should therefore be rejected itself as a piece of meaningless metaphysics.

---

[1] Shapere 1984a: xxvi.

[2] Putnam 1983e: 244.

On the other hand, my proposal does not endorse a purely descriptivist (and therefore anti-normative) conception of truth and objectivity. On the basis of my proposal, we can look at philosophy in general, and at positive philosophy in particular, as an integral part of cognitive discourse, interacting with its other parts and aiming at objective and rational validity, in the same way in which other intellectual constructions, such as scientific theories or certain forms of religion and of influential metaphysics, aspire to truth.

From my point of view, philosophy must operate in symbiosis with all the forms of human experience. It cannot live and develop in isolation, in the solitary search for the metaphysical or a priori foundations of knowledge and of behavioural norms. If it does not want to become sterile, degenerating into empty verbiage, philosophy must interact positively with the other forms of culture, operating within the aporias, the lacunae, the leaks, the obscurities of the ever changing conceptual and cultural system in which we are immersed. And philosophy's proposals of rational reconstruction are to be judged according to the same regulative ideals of rationality, objectivity and truth used to judge the more limited theories connected to other areas of cultural life.

The conception that I have suggested tries to realise this goal. It aims at avoiding, on the one hand, the contradictory character of radical relativism and, on the other hand, the metaphysical inflation of empirical realism which, with its notion of correspondence with a reality in itself, has no role in real cognitive activity, and condemns us to scepticism.

To be sure, the rejection of metaphysical absolutism will not be acceptable to those who believe that in knowledge, as in other spheres of cultural life (especially in ethics), it is not possible to do without absolute reference points. This objection carries little weight, for it is not sufficient to proclaim the existence of a metaphysical reference, in order to possess instruments of epistemological and ethical evaluation different from those that we all possess, and which depend always, and only, upon the best relative stopping points that we are able to devise.

I frankly admit that in my conception the alternative to relativism and absolutism has to be paid for by emptying of all content the fundamental notions of truth, objectivity and rationality. I have already explained why I do not think this is a defect. Contents depend upon the concrete problems faced in each particular instance; and the necessary parameters for the formulation and the solution of problems can emerge only from concrete cases. The general scheme must be empty and remain empty, in the same way as the principle of good will in Kant's moral philosophy should have been empty and should have remained empty. Of course, it should be kept firm that, in principle, the development of particular questions can always interact with the general scheme and make more or less deep modifications of it necessary. The rational construction of such a reference scheme constitutes the philosophical conclusion of this book. Nothing more – and nothing less.

BIBLIOGRAPHY

Achinstein, Peter (1968) *Concepts of Science. A Philosophical Analysis* (Baltimore: Johns Hopkins Press)

Allison, Henry E. (1983) *Kant's Transcendental Idealism. An Interpretation and Defence* (New Haven: Yale University Press)

Austin, John L. (1970) 'A Plea for Excuses' (1957), *in* J. Austin, *Philosophical Papers* (1961), J.O. Urmson and G.J. Warnock (eds), second edition (Oxford: Clarendon Press), pp. 175-204

Ayer, Alfred J. (1962) *Language, Truth and Logic* (1936; second edition revised, 1946), fifteenth impression of second edition (London: Gollancz)

Bain, Alexander (1870) *Logic*, 2 vols. (London: Longmans)

Beck, Lewis W. (1978) *Essays on Kant and Hume* (New Haven: Yale University Press)

Bernays, Paul (1974) 'Concerning Rationality', *in* Schilpp (ed) 1974: 597-605

Black, Max (1962) *Models and Metaphors. Studies in Language and Philosophy* (Ithaca: Cornell University Press)

Bolzano, Bernard (1972) *Theory of Science. Attempt at a Detailed and in the main Novel Exposition of Logic With Constant Attention to Earlier Authors* (*Wissenschaftslehre. Versuch einer ausführlichen und größtentheils neuen Darstellung der Logik mit steter Rücksicht auf deren bisherige Bearbeiter* (1837)), [abbreviated version] ed. / trans. R. George (Oxford: Blackwell) [References in square brackets are to the volume and page of the previous German editions (1837, 1882, 1914, 1929-31)]

Boyd, Richard N. (1984) 'The Current Status of Scientific Realism', *in* Leplin (ed) 1984: 41-82

Broad, Charlie D. (1978) *Kant. An Introduction*, C. Lewy (ed) (Cambridge: Cambridge University Press)

Brown, Harold I. (1979) *Perception, Theory and Commitment. The New Philosophy of Science* (1977) (Chicago: Phoenix edition)

(1987) *Observation and Objectivity* (New York / Oxford: Oxford University Press)

(1988) *Rationality* (London / New York: Routledge)

Buck, Roger C.; Cohen, Robert S. (eds) (1971) *Proceedings of the 1970 Biennial Meeting Philosophy of Science Association* "Boston Studies in the Philosophy of Science", vol. VIII, PSA 1970, In Memory of Rudolf Carnap (Dordrecht: Reidel)

Campbell, Norman R. (1957) *Foundations of Science. The Philosophy of Theory and Experiment*, unabridged and unaltered republication of the work formerly published under the title *Physics: the Elements*, (1920) (New York: Dover)

Carnap, Rudolf (1963a) 'Intellectual Autobiography', *in* Schilpp (ed) 1963: 3-84

(1963b) 'Replies and Systematic Expositions', *in* Schilpp (ed) 1963 859-1013

(1966) 'The Elimination of Metaphysics Through Logical Analysis of Language' (Überwindung der Metaphysik durch logische Analyse der Sprache) (1932), trans. A. Pap, *in* A.J. Ayer (ed), *Logical Positivism* (1959), second printing (New York: Free Press), pp. 60-81

(1967a) *The Logical Structure of the World* and *Pseudoproblems in Philosophy* (*Der Logische Aufbau der Welt* und *Scheinprobleme in der Philosophie* (1928)), trans. R.A. George (London: Routledge & Kegan Paul)

(1967b) 'Pseudoproblems in Philosophy' ('Scheinprobleme in der Philosophie' (1928)), *in* Carnap 1967a: 301-43

(1971) *The Logical Syntax of Language* (1937) (*Logische Syntax der Sprache* (1934)), trans. A. Smeaton (London: Routledge & Kegan Paul)

(1995) *An Introduction to the Philosophy of Science* (1966: original title, *Philosophical Foundations of Physics. An Introduction to the Philosophy of Science*), M. Gardner (ed) (New York: Dover)

Cassirer, Ernst (1922) *Das Erkenntnisproblem in der Philosophie und Wissenschaft der neueren Zeit*, II Vol., third edition (Berlin: Cassirer)

(1950) *The Problem of Knowledge. Philosophy, Science, and History Since Hegel* (*Das Erkenntnisproblem in der Philosophie und Wissenschaft der neueren Zeit.* Vierter Band: *Von Hegel's Tode bis zur Gegenwart. 1832-1932*), trans. W H. Woglom and C.W. Hendel (published prior to a German edition) (New Haven: Yale University Press)

(1953a) *Substance and Function* and *Einstein's Theory of Relativity* (1923) (*Substanzbegriff und Funktionsbegriff* (1910) and *Zur Einstein'schen Relativitätstheorie* (1921)), authorized trans. by W. Curtis Swabey and M. Collins Swabey (New York: Dover)

(1953b) *Einstein's Theory of Relativity Considered from the Epistemological Standpoint* (*Zur Einstein'schen Relativitätstheorie* (1921)), *in* Cassirer 1953a: 347-456

(1953c) *Language and Myth* (1946) (*Sprache und Mythos* (1925)), trans. S.K. Langer, New York: Dover

(1981) *Kant's Life and Thought* (*Kant's Leben und Lehre* (1918)) trans. J. Haden., Introduction by S. Körner (New Haven: Yale University Press)

Chisholm, Roderick M. (1957) *Perceiving: A Philosophical Study* (Ithaca: Cornell University Press)

Chomsky, Noam (1975) 'Quine's Empirical Assumptions', *in* Davidson and Hintikka (eds) 1975: 53-68

Clifford, William K. (1879) 'On the Nature of Things-in-Themselves' (1878), *in* W.K. Clifford, *Lectures and Essays*, L. Stephen and Pollock F. (eds), 2 vols., vol II, pp. 71-88

Cohen, L. Jonathan (1988) 'Realism Versus Anti-Realism: What Is the Issue?', *in* Ullmann-Margalit (ed) 1988: 81-96

(1989) *An Introduction to the Philosophy of Induction and Probability* (Oxford: Clarendon)

Davidson, Donald (1984) 'On the Very Idea of a Conceptual Scheme' (1974), *in* D. Davidson, *Inquiries into Truth and Interpretation* (Oxford: Clarendon Press), pp. 183-98

Davidson, Donald; Hintikka, Jaakko (eds) (1975) *Words and Objections. Essays on the Work of W.V. Quine* (1969) (Dordrecht: Reidel)

Devitt, Michael (1984) *Realism and Truth* (Oxford: Blackwell)

Duhem, Pierre (1962a) *The Aim and Structure of Physical Theory (La Théorie Physique: Son Objet, Sa Structure* (1904-1906, 1914)), trans. P.P. Wiener, from the second edition (1914) (New York: Atheneum)

(1962b) 'The Value of Physical Theory. Concerning a Recent Book [Rey 1907]' ('La Valeur de la Théorie Physique. A propos d'un livre récent' (1908)), *in* Duhem 1962a: 312-35

Earman, John (1992) *Bayes or Bust? A Critical Examination of Bayesian Confirmation Theory* (Cambridge (Mass.): The MIT Press)

Earman, John; Salmon, Wesley C. (1992) 'The Confirmation of Scientific Hypotheses', *in* M Salmon (et al) 1992: 42-103

Einstein, Albert (1951) 'Reply to Criticisms', *in* Paul A. Schilpp (ed), *Albert Einstein: Philosopher-Scientist* (1949), second edition (New York: Tudor), pp. 663-88

Enriques, Federigo (1936) 'Philosophie scientifique', *in Actes du Congrès International de Philosophie Scientifique, Sorbonne, Paris, 1935*, II: *Rationalisme empirique et empirisme logique* (Paris: Hermann), pp. 23-7

Feigl, Herbert (1979) 'Moritz Schlick, A Memoir' (1938), *in* Schlick 1979a: xv-xxxviii
Feyerabend, Paul K. (1965) 'Problems of Empiricism I', *in* Robert G. Colodny (ed), *Beyond the Edge of Certainty. Essays in Contemporary Science and Philosophy* (Englewood Cliffs: Prentice Hall), pp. 145-260
(1973) *Contro il metodo (Against Method. Outline of an anarchistic theory of knowledge*), Italian trans. (Milano: Lampugnani Nigri)
(1979) *Against Method. Outline of an anarchistic theory of knowledge* (first Verso edition, 1975) (London: Verso)
(1981a) *Realism, Rationalism and Scientific Method. Philosophical Papers, Vol I* (Cambridge: Cambridge University Press)
(1981b) 'An attempt at a realistic interpretation of experience' ('Zur Theorie der Basissätze' (1958)), *in* Feyerabend 1981a: 17-36
(1981c) 'Explanation, Reduction and Empiricism' (1962), *in* Feyerabend 1981a: 44-96
(1983) *Il realismo scientifico e l'autorità della scienza (Der wissenschaftstheoretische Realismus und die Autorität der Wissenschaften* (1978)) Italian trans., A. Artosi (ed) (Milano: Il Saggiatore)
(1987) 'Consolations for the Specialist'(1970), *in* Lakatos and Musgrave (eds) 1987: 197-230
Fine, Arthur (1984) 'The Natural Ontological Attitude', *in* Leplin (ed) 1984: 83-107
Fine, Arthur; Forbes, M. (1986) 'Grünbaum on Freud: Three grounds for dissent', *The Behavioral and Brain Sciences* 9, pp. 237-8
Fleck, Ludwik (1979) *Genesis and Development of a Scientific Fact (Entstehung und Entwicklung einer wissenschaftlichen Tatsache. Einführung in die Lehre vom Denkstil und Denkkollektiv)* (1935), T.J. Trenn and R K. Merton (eds), trans. F. Bradley and T.J. Trenn (Chicago / London: The University of Chicago Press)
Frank, Philipp (1975) *Modern Science and Its Philosophy* (1941, 1949) (New York: Arno Press)
Franklin, Allan (1986) *The Neglect of Experiment* (Cambridge: Cambridge University Press)
Gibson, Roger F., Jr. (1986) 'Translation, Physics, and Facts of the Matter', *in* L.E. Hahn.; P.A. Schilpp (eds) *The Philosophy of W.V. Quine* (La Salle: Open Court), pp. 139-54
Glymour, Clark (1980) *Theory and Evidence* (Princeton: Princeton University Press)
Gödel, Kurt (1990) 'Russell's Mathematical Logic' (1944), *in* K. Gödel, *Collected Works, vol. II (publications 1938-1974)* (S. Feferman et al., eds.) (New York / Oxford: Oxford University Press), pp. 102-41
Goodman, Nelson (1978) 'Words, Works, Worlds' (1975), *in* N. Goodman, *Ways of Worldmaking* (Hassocks: Harvester Press), pp. 1-22
(1983) 'Prospects for a Theory of Projection' (1954) *in* N. Goodman, *Fact, Fiction, and Forecast* (1954), fourth edition (Cambridge (Mass.): Harvard University Press), pp. 84-124
Goodman, Nelson; Elgin, Catherine Z. (1988) 'Interpretation and Identity: Can the Work Survive the World?' (earlier version: 1986), *in* N. Goodman and C. Elgin *Reconceptions in Philosophy and Other Arts and Sciences* (London: Routledge), pp. 49-65
Grünbaum, Adolf (1973) *Philosophical Problems of Space and Time* (1963), second enlarged edition (Dordrecht: Reidel)
(1976) 'Is the Method of Bold Conjectures and Attempted Refutations *Justifiably* the Method of Science?', *British Journal of Philosophy of Science* **27**, pp 105-36
Guyer, Paul (1987) *Kant and the Claims of Knowledge* (Cambridge: Cambridge University Press)
Hacking, Ian (1983) *Representing and Intervening. Introductory Topics in the Philosophy of Natural Science* (Cambridge: Cambridge University Press)

(1984) 'Experimentation and Scientific Realism' (1982), *in* Leplin (ed) 1984: 154-72

Hahn, Hans (1987) 'Logic, Mathematics, and Knowledge of Nature' ('Logik, Mathematik und Naturerkenntnis' (1933)), *in* B. McGuinness (ed) *Unified Science. The Vienna Circle Monograph Series, originally edited by Otto Neurath, (Einheitswissenschaft* (1933-1939)), *now in an English edition. With an Introduction by R. Hegselmann; Translations by H. Kaal,* Vienna Circle Collection, Vol. 19 (Dordrecht. Reidel (Kluwer)), pp. 24-45

Hanson, Norwood R (1965) *Patterns of Discovery. An Inquiry into the Conceptual Foundations of Science* (1958) (Cambridge: Cambridge University Press)

Hartmann, Nicolai (1933) *Das Problem des Geistigen Seins* (Berlin: de Gruyter)

Hempel, Carl G. (1966) *Philosophy of Natural Science* (Englewood Cliffs (N.J): Prentice-Hall)

(1988) 'Limits of a Deductive Construal of the Function of Scientific Theories', *in* Ullmann-Margalit (ed) 1988:1-15

(1992) 'Eino Kaila and Logical Empiricism', *in* I. Niiniluoto, M. Sintonen, G H. Von Wright (eds), *Eino Kaila and Logical Empiricism*, 'Acta Philosophica Fennica' **52**, 1992: 43-51

Herbart, Johann F. (1964) *Lehrbuch zur Einleitung in die Philosophie, in* J.F. Herbart, *Sämtliche Werke* vol. 4, K. Kerback and O. Flügel (eds) (Darmstadt: Scientia Verlag Aalen)

Hertz, Heinrich (1956) *The Principles of Mechanics, Presented in a New Form* (1899) (*Die Prinzipien der Mechanik in neuem Zusammenhange dargestellt* (1894)), Preface by H. von Helmholtz; authorised English trans. by D.E. Jones and J.T. Walley; with a new Introduction by R.S. Cohen (New York: Dover)

Hesse, Mary B. (1974) *The Structure of Scientific Inference* (London: Macmillan)

(1976) 'Duhem, Quine and a New Empiricism' (1970), *in* S.G. Harding (ed) *Can Theories be Refuted? Essays on the Duhem-Quine Thesis* (Dordrecht: Reidel), pp. 184-204

(1980) *Revolutions and Reconstructions in the Philosophy of Science* (Brighton: Harvester)

Hoyningen-Huene, Paul (1993) *Reconstructing Scientific Revolutions. Thomas S. Kuhn's Philosophy of Science* (*Die Wissenschaftsphilosophie Thomas S. Kuhns: Rekonstruktion und Grundlagenprobleme*, Braunschweig: Vieweg, 1989)), trans. A.T. Levine, with a foreword by T.S. Kuhn (Chicago / London: University of Chicago Press)

Husserl, Edmund (1987) *Vorlesungen über Bedeutungslehre. Sommersemester 1908*, U. Panzer (ed), Husserliana. Edmund Husserl, *Gesammelte Werke*, Vol. XXVI (Dordrecht: Nijhoff (Kluwer))

(1990) *The Idea of Phenomenology* (1964) (*Die Idee der Phänomenologie. Fünf Vorlesungen* (lectures held in 1907; first German edition, 1950)), trans. W.P. Alston and G. Nakhnikian. Introduction by G. Nakhnikian. Sixth impression (Dordrecht: Kluwer) [References in square brackets are to the pages in the original German edition, *Husserliana* – Edmund Husserl, *Gesammelte Werke.* Auf Grund des Nachlasses veröffentlicht mit dem Husserl-Archiv an der Universität Köln vom Husserl Archiv (Louvain) unter Leitung von H L. Van Breda, Band II: *Die Idee der Phänomenologie. Fünf Vorlesungen*, herausgegeben von W. Biemel, Haag, Martinus Nijhoff, 1950]

Hutten, Ernest H. (1976) *Einstein e Freud. Creatività nella scienza.* (*Creative Science: Einstein and Freud (An Interdisciplinary Study)* (1976)), Italian trans. by A. Artani (Roma: Armando)

Johnson, Lawrence E. (1992) *Focusing on Truth* (London: Routledge)

Kant, Immanuel (1985) *Critique of Pure Reason* (1929; second impression with corrections, 1933) (*Kritik der Reinen Vernunft* (1781, 1787)), trans. N. Kemp Smith

(London: Macmillan) [References in square brackets are to the first (A) and second (B) German edition (Riga: Hartknoch, 1781 and 1787)]

(1992) *Logic. A Manual for Lectures* (*Logik*, G.B. Jäsche (ed) (1800)), *in* I. Kant, *Lectures on Logic*, Trans / ed. J. M. Young. (The Cambridge Edition of the Works of Immanuel Kant) (Cambridge: Cambridge University Press), pp. 521-640 [References in square brackets are to the volume and page of the Akademie edition (*Kant's Gesammelte Schriften*. Herausgegeben von der Preussischen Akademie der Wissenschaften. Vol 9 (1956))]

(1993) *Opus Postumum*, E. Förster (ed), trans. for the first time into English by E. Förster and M. Rosen (Cambridge: Cambridge University Press) [References in square brackets are to the volume and page of the Akademie edition, Vols. 21 (1936) and 22 (1938)]

(1995) *Prolegomena To Any Future Metaphysics That Will Be Able To Come Forward As Science*. (*Prolegomena zu einer jeden künftigen Metaphysik, die als Wissenschaft wird auftreten können* (1783)). The Paul Carus Translation extensively revised by J.W. Ellington (1977) (Indianapolis / Cambridge: Hackett) [References in square brackets are to the Akademie edition, Vol. 4]

Katz, Jerrold J. (1967) 'Some Remarks on Quine on Analyticity', *Journal of Philosophy (64)*, 1967, pp 36-52

(1968) 'Unpalatable Recipes for Buttering Parsnips', *Journal of Philosophy (65)*, 1968, pp 29-45

Kemp Smith, Norman (1979) *A commentary to Kant's 'Critique of Pure Reason'* (1918) (London: Macmillan)

Koertge, Noretta (1980) 'Analysis as a Method of Discovery during the Scientific Revolution', *in* Nickles (ed) 1980a: 139-57

Kordig, Carl R. (1971) *The Justification of Scientific Change* (Dordrecht: Reidel)

Körner, Stephan (1974) *Categorical Frameworks* (1970) (Oxford: Blackwell)

Kraft, Victor (1961) 'Développement et évolution du Positivisme', *in* M. Sciacca, *Les Grands Courants de la Pensée Mondiale Contemporaine*, vol. II, Milano: Marzorati : 817-70

Kuhn, Thomas S. (1971) 'Notes on Lakatos', *in* Buck and Cohen (eds): 137-46

(1977a) *The Essential Tension. Selected Studies in Scientific Tradition and Change* (Chicago; University of Chicago Press)

(1977b) 'Objectivity, Value Judgment, and Theory Choice' (previously unpublished lecture (1973)), *in* Kuhn 1977a: 320-339

(1977c) 'Second Thoughts on Paradigms' (1974), *in* Kuhn 1977a: 293-319

(1983) 'Rationality and Theory Choice', *The Journal of Philosophy* **80**, n. 10, pp. 563-70

(1987a) 'Logic of Discovery or Psychology of Research?' (1970) *in* Lakatos and Musgrave (eds) 1987: 1-24

(1987b) 'Reflections on My Critics' (1970), *in* Lakatos and Musgrave (eds) 1987: 231-78

(1996a) *The Structure of Scientific Revolutions* (1962), Third Edition (Chicago; London: University of Chicago Press)

(1996b) 'Postscript 1969' *in* Kuhn 1996a: 174-210

Lakatos, Imre (1971) 'History of Science and its Rational Reconstructions' (1970), *in* Buck and Cohen (eds) 1971: 91-136

(1987) 'Falsification and the Methodology of Scientific Research Programmes' (improved version of a paper published in 1968), *in* Lakatos and Musgrave (eds) 1987: 91-196

Lakatos, Imre; Musgrave, Alan (eds) (1987) *Criticism and the Growth of Knowledge. Proceedings of the International Colloquium in the Philosophy of Science, London 1965, volume 4* (1970) (Cambridge: Cambridge University Press)

Lanfredini, Roberta (1988) *Oggetti e paradigmi. Per una concezione interattiva della conoscenza scientifica* (Roma: Theoria)

Laudan, Larry (1981) *Science and Hypothesis. Historical Essays on Scientific Methodology* (Dordrecht: Reidel)

(1984) 'A Confutation of Convergent Realism' (1981), *in* Leplin (ed) 1984: 218-49

Leibniz, Gottfried W. (1965) 'Schreiben Leibnizens in Betreff der Schrift Joh. Eberh. Schweiling's zur Vertheidigung des Cartesianismus. Eine kurze Abhandlung Leibnizens von demselben Inhalt', *in Die philosophischen Schriften*, vol IV, C.I Gerhardt (ed) (Hildesheim: Olms), pp. 325-32

Leplin, Jarrett (1984) Introduction to Leplin (ed) 1984· pp. 1-7

Leplin, Jarrett (ed) (1984) *Scientific Realism* (Berkley / Los Angeles: University of California Press)

Lewis, Clarence I. (1956) *Mind and the World-Order. Outline of a Theory of Knowledge* (1929), revised edition (New York: Dover)

Longino, Helen L. (1979) 'Evidence and Hypothesis: An Analysis of Evidential Relations', *Philosophy of Science* **46**, pp. 35-56

Lyotard, Jean-François (1992) *Peregrinazioni. Legge, forma, evento* (*Peregrinations. Law, Form, Event* (1988)) (Bologna: Il Mulino)

Mach, Ernst (1911) *History and Root of the Principle of the Conservation of Energy* (*Die Geschichte und die Wurzel des Satzes von der Erhaltung der Arbeit* (1871)) (Chicago: Open Court)

(1959) *The Analysis of Sensation and the Relation of the Physical to the Psychical (Die Analyse der Empfindungen und das Verhältnis des Physischen zum Psychischen* (1886)). With a New Introduction by T.S Szasz, Translated from the first German edition by C M. Williams; revised and supplemented from the fifth German edition by S. Waterlow (New York: Dover)

(1974) *The Science of Mechanics: A Critical & Historical Account of Its Development* (*Die Mechanik in Ihrer Entwicklung Historisch-Kritisch Dargestellt* (1883)), Sixth edition, with revisions through the ninth German edition (La Salle (Ill.): Open Court)

(1986) 'On the Economical Nature of Physical Inquiry' ('Die ökonomische Natur der physikalischen Forschung'), *in* E. Mach, *Popular Scientific Lectures* (1893. Third edition, with additional lectures, 1898. Fifth edition 1943. First paperback edition) (*Popularwissenschaftliche Vorlesungen* (1895)) (La Salle (Ill.): Open Court), pp. 186-213

Machamer, Peter (1995) 'Kitcher and the Achievement of Science', *Philosophy and Phenomenological Research* **55**, n. 3, pp. 629-36

Martin, Michael (1971) 'Referential Variance and Scientific Objectivity', *British Journal for the Philosophy of Science* **22**, pp 17-26

(1972) 'Ontological Variance and Scientific Objectivity', *British Journal for the Philosophy of Science* **23**, pp 252-256

Mathieu, Vittorio (1958) *La filosofia trascendentale e l' 'Opus Postumum' di Kant* (Torino. Edizioni di Filosofia)

McMullin, Ernan (1984) 'A Case for Scientific Realism' *in* Leplin (ed) 1984: 8-40

Neurath, Otto (1935) *Le développement du Cercle de Vienne et l'avenir de l'empirisme logique* (1935) traduction du Général Vouillemin, 'Actualités Scientifiques et Industrielles, 290' (Paris: Hermann)

(1973) *Empiricism and Sociology*, M. Neurath and R.S. Cohen (eds) (Dordrecht: Reidel)

(1983) 'Radical Physicalism and the 'Real World' ('Radikaler Physicalismus und 'wirkliche welt'', *Erkenntnis*, 4, 1934: 346-62), *in* O. Neurath, *Philosophical Papers 1913-1946*, R.S. Cohen and M. Neurath (eds), "Vienna Circle Collection", Dordrecht: Reidel (Kluwer), pp 100-14

Neurath, Otto; Hahn, Hans; Carnap, Rudolf (1973) *Wissenschaftliche Weltauffassung: der Wiener Kreis* (1929) *[The Scientific Conception of the World: the Vienna Circle]*, English trans. *in* Neurath 1973: 299-318

Nickles, Thomas (ed) (1980a) *Scientific Discovery, Logic, and Rationality*, Boston Studies in the Philosophy of Science, Vol. 56 (Dordrecht: Reidel)

(1980b) *Scientific Discovery: Case Studies*, Boston Studies in the Philosophy of Science, Vol. 60 (Dordrecht: Reidel)

Nozick, Robert (1981) *Philosophical Explanations* (Cambridge (Mass.): Harvard University Press)

Parrini, Paolo (1976a) *Linguaggio e teoria. Due saggi di analisi filosofica*, Firenze: La Nuova Italia

(1976b) 'L'analiticità e il problema del significato' (1972-3), *in* Parrini, 1976a. 19-151

(1976c) 'La logica del controllo empirico e il problema dell'a priori', *in* Parrini 1976a:153-290

(1980a) *Una filosofia senza dogmi. Materiali per un bilancio dell'empirismo contemporaneo* (Bologna: Il Mulino)

(1980b) 'Per un bilancio dell'empirismo contemporaneo: contributo alla storia del positivismo logico' (an essay that developed from a discussion at the XXIV Congresso Nazionale di Filosofia, *Bilancio dell'empirismo contemporaneo*, L'Aquila, 28 aprile - 2 maggio 1973, printed in *Rivista di storia della filosofia*, XXXI, 1976: 193-239), reprinted with modifications and additions in Parrini 1980a: 21-92

(1980c) 'La struttura logica del controllo empirico e il falsificazionismo popperiano', *in* Parrini 1980a: 93-162

(1980d) 'Linguistica generativa, comportamentismo, empirismo' (1977), *in* Parrini 1980a:163-76

(1983) *Empirismo logico e convenzionalismo* (Milano: Angeli)

(1987) 'Empirismo logico, kantismo e convenzionalismo' *Paradigmi* **5**, pp. 243-61

(1988a) 'Preti teorico della conoscenza' (1987), *Annali del Dipartimento di Filosofia dell'Università di Firenze* **4**, pp. 131-65

(1988b) 'La psicoanalisi nella filosofia della scienza', *Rivista di Storia della Filosofia* **43**, pp. 165-80

(1993) 'Origini e sviluppi dell'empirismo logico nei suoi rapporti con la filosofia continentale. Alcuni testi inediti [di Schlick, di Reichenbach, di Carnap]', *Rivista di storia della filosofia* **48**, n. 1, pp 121-46; n. 2, pp 377-93

(1994a) 'On Kant's Theory of Knowledge: Truth, Form, Matter' (1990), *in* Parrini (ed) 1994: 195-230

(1994b) 'With Carnap Beyond Carnap. Metaphysics, Science and the Realism/Instrumentalism Controversy' (1991), *in* W.C. Salmon; G. Wolters (eds) *Logic, Language, and the Structure of Scientific Theories. Proceedings of the Carnap-Reichenbach Centennial, University of Konstanz, 21-24 May 1991* (Pittsburgh / Konstanz: University of Pittsburgh Press and Universitätsverlag Konstanz), pp. 255-77

(1994c) 'Reichenbach, la teoria della relatività e la problematica dell'a priori' (1990), *in* S. Petruccioli (ed), *Dagli atomi di elettricità alle particelle atomiche. Problemi di storia e filosofia della fisica tra Ottocento e Novecento*, "Lezioni Galileiane", vol. IV (Roma: Istituto dell'Enciclopedia Italiana), pp. 223-47

(199+a) ''Immanenzgedanken' and Knowledge as Unification. Scientific Philosophy and Philosophy of Science' (1996), talk given at a workshop on 'Science et philosophie en France et en Italie entre les deux guerres', Paris, 10-11 May 1996, and at a workshop in Honour of Merrilee and Wesley Salmon, 'Experience, Reality, and Scientific Explanation', Florence, 17-18 May, 1996 (the proceedings of the workshop are forthcoming in the University of Western Ontario Series in the Philosophy of Science)

(199+b) 'Realismo, scetticismo e analisi filosofica'; forthcoming

Parrini, Paolo (ed) (1994) *Kant and Contemporary Epistemology*, The University of Western Ontario Series in Philosophy of Science. Papers presented at an international workshop held at the Florence Centre for the History and Philosophy of Science, May 27-30, 1992 (Dordrecht: Kluwer)

Peirce, Charles S. (1960) 'The Critic of Arguments' (1892), *in* C.S. Peirce, *Collected Papers*, C Hartshorne and P. Weiss (eds), Vol. III, *Exact Logic* (Cambridge (Mass): The Belknap Press of Harvard University Press), pp. 250-65 [References in square brackets are to the volume and paragraph of Peirce's *Collected Papers*]

Pippin, Robert B. (1982) *Kant's Theory of Form. An Essay on the 'Critique of Pure Reason'* (New Haven: Yale University Press)

(1986) Review of Allison 1983, *Kant Studien*, (77), 1986: 365-71

Plato (1914) *Phaedo*, in *Plato, with an English translation by H.N. Fowler and an Introduction by W.R.M. Lamb*, vol. I. (The Loeb Classical Library, T.E. Page and W H D. Rouse eds.) (London: Heinemann / New York: Macmillan), pp. 193-403. [References in square brackets are to the volume and page number of the edition of H. Stephanus (Paris, 1578)]

Poincaré, Henri (1900) 'Sur les Principes de la Géométrie. Réponse a M. Russell', *Revue de Métaphysique et de Morale* **8**, pp 73-86

(1946a) *The Foundations of Science* [*Science and Hypothesis; The Value of Science; Science and Method*] Authorized translation by G.B. Halsted, With a special Preface by Poincaré, and an Introduction by J. Royce (1913) (Lancaster (Pa.): The Science Press)

(1946b) 'Science and Hypothesis' ('La Science et l'hypothèse' (1902)), *in* Poincaré 1946a: 1-197

(1946c) 'The Value of Science' ('La valeur de la science' (1905)), *in* Poincaré 1946a: 199-355

Polanyi, Michael (1958) *Personal Knowledge. Towards a Post-Critical Philosophy* (London: Routledge & Kegan Paul)

(1966) *The Tacit Dimension* (London: Routledge & Kegan Paul)

Popper, Karl R. (1972a) *Conjectures and Refutations. The Growth of Scientific Knowledge* (1963), fourth edition (revised) (London: Routledge and Kegan Paul)

(1972b) 'Truth, Rationality and the Growth of Scientific Knowledge' (written in 1960), *in* Popper 1972a: 215-50

(1974a) 'Replies to my Critics', *in* Schilpp 1974: 959-1197

(1974b) 'Lakatos on the Equal Status of Newton's and Freud's Theories', *in* Popper 1974a: 999-1013

(1974c) 'Bernays's Plea for a Wider Notion of Reality', *in* Popper 1974a: 1081-91

(1974d) 'Ayer on Empiricism and Against Verisimilitude', *in* Popper 1974a: 1100-14

(1975a) *Objective Knowledge. An Evolutionary Approach* (1972) reprinted (with corrections) (Oxford: Clarendon Press)

(1975b) 'Two Faces of Commonsense: An Argument for Commonsense Realism and Against the Commonsense Theory of Knowledge' (revised and expanded version of a talk given in 1970), *in* Popper 1975a: 32-105

(1977a) *The Logic of Scientific Discovery* (*Logik der Forschung* (1934)), translated by K.R. Popper with the assistance of J. and L. Freed (1959). London: Hutchinson

(1977b) New Appendices (1959), *in* Popper 1977a: 311-464

(1983) *Realism and the Aim of Science*. From the *Postcript to the Logic of Scientific Discovery*, vol I (1956), W.W. Bartley III (ed) (London: Hutchinson)

(1986) 'How I see Philosophy' (revised version of a paper first published in 1975), *in* S. G Shanker (ed) *Philosophy in Britain Today*, London: Croom Helm, pp. 198-212

(1987) 'Normal Science and Its Dangers' (1970), *in* Lakatos and Musgrave (eds) 1987: 51-8

Preti, Giulio (1940) 'Il significato metafisico del problema della conoscenza', *Studi filosofici* **1**, pp. 26-43; 230-9
(1943) *Idealismo e positivismo* (Milano. Bompiani)
(1946) 'I limiti del neopositivismo', *Studi Filosofici* **7** [**6**], pp. 87-96
(1957) *Praxis ed empirismo* (Torino. Einaudi)
(1968) *Retorica e logica* (Torino: Einaudi)
(1974) 'Lo scetticismo e il problema della conoscenza', *Rivista [critica] di storia della filosofia* **29**, pp. 3-31; 123-43, 243-63
(1976a) *Saggi filosofici*, Mario Dal Pra (ed), Vol. I: *Empirismo logico, epistemologia e logica* (Firenze: La Nuova Italia)
(1976b) 'Linguaggio comune e linguaggi scientifici' (1953), *in* Preti 1976a. 127-220
(1976c) 'L'ontologia della regione 'natura' nella fisica newtoniana' (1957), *in* Preti 1976a: 413-3
(1976d) 'Il mio punto di vista empiristico' (1958), *in* Preti 1976a: 475-95
(1976e) 'Il linguaggio della filosofia' (1962), *in* Preti 1976a: 455-7
(1976f) 'Pluralità delle scienze e unità eidetica del mondo scientifico' (1965), *in* Preti 1976a: 497-512
Putnam, Hilary (1975a) *Mind, Language and Reality. Philosophical Papers, vol II* (Cambridge: Cambridge University Press)
(1975b) 'The Analytic and the Synthetic' (1962), *in* Putnam 1975a: 33-69
(1975c) 'How Not to Talk about Meaning. Comments on J.J.C. Smart' (1965), *in* Putnam 1975a: 117-31
(1979) *Meaning and the Moral Sciences* (1978) (Boston / London / Henley: Routledge & Kegan Paul)
(1982) *Reason, Truth and History* (Cambridge: Cambridge University Press)
(1983a) *Realism and Reason, Philosophical Papers vol III* (Cambridge: Cambridge University Press)
(1983b) 'Models and Reality' (1977), *in* Putnam 1983a: 1-25
(1983c) 'Reflections on Goodman's *Ways of Worldmaking*' (1979), *in* Putnam 1983a: 155-69
(1983d) 'Beyond Historicism'(1981), *in* Putnam 1983a: 287-303
(1983e) 'Why Reason Can't Be Naturalized' (1981), *in* Putnam 1983a: 229-47
(1983f) 'Why There Isn't a Ready-made World' (1981), *in* Putnam 1983a: 205-28
(1988) *Representation and Reality* (Cambridge (Mass.). The MIT Press)
Quine, Willard V. O. (1964a) *From a Logical Point of View: Nine Logico-Philosophical Essays* (1953), second edition, revised (Cambridge (Mass.): Harvard University Press)
(1964b) 'New Foundations for Mathematical Logic' (1937), *in* Quine 1964a: 80-101
(1964c) 'Identity, Ostension and Hypostasis' (1950), *in* Quine 1964a: 65-79
(1964d) 'Two Dogmas of Empiricism' (1951), *in* Quine 1964a: 20-46
(1967) 'On a Suggestion of Katz', *Journal of Philosophy* **64**, pp 52-4
(1970) *Word and Object* (1960) (Cambridge (Mass.): The MIT Press)
(1975a) 'Reply to Chomsky' (1969), *in* Davidson and Hintikka (eds) 1975: 302-11
(1975b) 'On Empirically Equivalent Systems of the World', *Erkenntnis* **9**, pp 313-28
(1977) 'Carnap and Logical Truth' (1954), *in* W.V.O. Quine *The Ways of Paradox and Other Essays* (1966), Revised and enlarged edition (1976), Second printing (Cambridge (Mass.): Harvard University Edition), pp. 107-32
(1985) *The Time of My Life. An Autobiography* (Cambridge (Mass.) / London. The MIT Press)
(1986) *Philosophy of Logic* (1970), second edition (Cambridge (Mass.): Harvard University Press)
(1990) *Pursuit of Truth* (Cambridge (Mass.): Harvard University Press)

(1995) 'Whitehead and the Rise of Modern Logic' (1941) *in* W.V O. Quine *Selected Logic Papers* (1966), Enlarged edition (Cambridge (Mass.): Harvard University Press), pp. 3-36

Reichenbach, Hans (1931) Letters published in 'Philosophy of science' **6**, pp. 398-9; 525-6

(1958) *The Philosophy of Space and Time* (*Philosophie der Raum-Zeit-Lehre* (1928)), trans M. Reichenbach and J. Freund, with Introductory remarks by R Carnap (New York: Dover)

(1965) *The Theory of Relativity and A Priori Knowledge* (*Relativitätstheorie und Erkenntnis Apriori* (1920)), ed. / trans. M. Reichenbach (Berkley / Los Angeles: University of California Press)

(1978a) *Selected Writings*, Vol. I. M. Reichenbach & R.S. Cohen, eds (Dordrecht: Reidel)

(1978b) *Selected Writings,* Vol II, M Reichenbach & R S. Cohen, eds (Dordrecht: Reidel)

(1978c) 'The Philosophical Significance of Modern Physics' ('Die Philosophische Bedeutung der Modernen Physik' (1930)), *in* Reichenbach 1978a: 304-23

(1978d) 'Kant and Natural Science' ('Kant und die Naturwissenschaft' (1933)), *in* Reichenbach 1978a: 389-404

(1978e) 'The Present State of the Discussion on Relativity. A Critical Investigation' ('Der gegenwärtige Stand der Relativitätsdiskussion' (1922)), *in* Reichenbach 1978b. 3-47

(1978f) 'The Aims and Method of Physical Knowledge' ('Ziele und Wege der physikalische Erkenntnis' (1929)), *in* Reichenbach 1978b: 120-225

Rescher, Nicholas (1987) *Scientific Realism. A Critical Reappraisal* (Dordrecht: Reidel (Kluwer))

Rey, Abel (1907) *La Théorie de la physique chez les physiciens contemporains* (Paris: Alcan)

Rorty;, Richard (1980) *Philosophy and the Mirror of Nature* (Oxford: Blackwell)

(1982) 'The World Well Lost' (1972), *in* R Rorty;, *Consequences of Pragmatism (Essays 1972-1980)* (Brighton: Harvester), pp. 3-18

Russell, Bertrand (1950) *Introduction to Mathematical Philosophy* (1919) (London: Allen and Unwin)

(1975) *My Philosophical Development* (1959), second edition (London: Unwin Books edition, Allen & Unwin)

(1990) 'The Axioms of Geometry' ('Sur les Axiomes de la Géométrie' (1899)), *in* B. Russell *Philosophical Papers 1896-99*, N. Griffin and A.C. Lewis, eds; Textual Apparatus Prepared by W.G. Stratton (London: Unwin Hyman), pp. 390-415; French text on pp. 432-51

Salmon, Merrilee *et al.* (ed) (1992) *Introduction to the Philosophy of Science. A Text by Members of the Department of the History and Philosophy of Science of the University of Pittsburgh* (Englewood Cliffs (N.J.): Prentice Hall)

Salmon, Wesley C. (1988) 'Rational Prediction' (a version of this paper presented orally in 1980), *in* A. Grünbaum and W. Salmon (eds) *The Limitations of Deductivism* (Berkeley / Los Angeles: University of California Press), pp. 47-60

(1992) 'Scientific Explanation', *in* M. Salmon et al (eds) 1992: 7-41

(1993) 'The Value of Scientific Understanding', *Philosophica* **51**, n. 1, pp. 9-19

Santayana, George (1962) *Reason in Common Sense*. Vol. I of *The Life of Reason* (1909, 1922) (New York: Collier Books)

Sapir, Edward (1951) 'The Status of Linguistics as a Science' (1929, *in* E. Sapir, *Selected Writings of Edward Sapir in Language, Culture and Personality* (1949), D.G. Mandelbaum (ed) (Berkeley, Los Angeles, London: University of California Press), pp. 160-6

(1982) *Language. An Introduction to the Study of Speech* (1921) (London / Toronto / Sydney / New York : Granada)

Scheffler, Israel (1976) *Science and Subjectivity* (1967) (Indianapolis: Bobbs-Merrill)

Schilpp, Paul A. (ed) (1963) *The Philosophy of R. Carnap* (La Salle (Ill.): Open Court)

(1974) *The Philosophy of K.R.* Popper (Books One & Two) (La Salle (Ill): Open Court)

Schlick, Moritz (1974) *General Theory of Knowledge* (*Allgemeine Erkenntnislehre* (1918, second revised edition, 1925)) trans. A. E. Blumberg, with an Introduction by A E. Blumberg and H Feigl (Wien / New York: Springer-Verlag)

(1978) 'Critical or Empiricist Interpretation of Modern Physics? Remarks on E Cassirer's *Einstein's Theory of Relativity*' [Cassirer 1953b] ('Kritizistische oder empiristische Deutung der neuen Physik?' (1921)), *in* M. Schlick, *Philosophical Papers*. vol. I (1909-1922), H L. Mulder and B.F.B. Van De Velde-Schlick (eds), "Vienna Circle Collection" (Dordrecht: Reidel (Kluwer)), pp. 322-34

(1979a) *Philosophical Papers*. vol. II (1925-36), H.L Mulder and B.F.B. Van De Velde-Schlick (eds), "Vienna Circle Collection" (Dordrecht: Reidel (Kluwer))

(1979b) 'Is There a Factual *a Priori*?' ('Gibt es ein materiales Apriori?' (1932)), *in* Schlick 1979a: 161-70

(1979c) 'Positivism and Realism' ('Positivismus und Realismus' (1932)), *in* Schlick 1979a: 259-84

(1979d) 'On the Foundation of Knowledge' ('Über das Fundament der Erkenntnis' (1934), *in* Schlick 1979a: 370-87

(1979e) 'Introduction' and 'On "Affirmations"' ('Introduction' and 'Sur les "Constatations"' (1935), *in* Schlick 1979a: 405-13

(1979f) 'The Vienna School and Traditional Philosophy' ('L'Ecole de Vienne et la Philosophie Traditionelle' (1937), *in* Schlick 1979a: 491-98

(1979g) 'Form and Content. An Introduction to Philosophical Thinking' (1938), *in* Schlick 1979a: 285-369

Sellars, Wilfrid (1967) 'Counterfactuals, Dispositions, and the Causal Modalities', *in* H. Feigl, M. Scriven, G. Maxwell (eds) *Minnesota Studies in the Philosophy of Science*, Vol II: *Concepts, Theories, and the Mind-Body Problem* (1958), Third edition (Minneapolis: University of Minnesota Press), pp. 225-308

(1968) *Science, Perception and Reality* (1963) (London; Routledge & Kegan Paul)

Shapere, Dudley (1984a) *Reason and the Search for Knowledge. Investigations in the Philosophy of Science* (Dordrecht: Reidel (Kluwer))

(1984b) 'Meaning and Scientific Change' (1966), *in* Shapere 1984a:58-101

Siegel, Harvey (1987) *Relativism Refuted: A Critique of Contemporary Epistemological Relativism* (Dordrecht: Kluwer)

Smith, Peter (1981) *Realism and the Progress of Science* (Cambridge: Cambridge University Press)

Solomon, Robert C. (1988) 'Continental Philosophy since 1750', vol VII of *A History of Western Philosophy* (Oxford: Oxford University Press)

Strawson, Peter F. (1966) *The Bounds of Sense. An Essay on Kant's 'Critique of Pure Reason'* (London: Methuen)

Suppe, Frederick (1977a) 'The Search for Philosophic Understanding of Scientific Theories' (1973), *in* Suppe (ed) 1977: 1-241

(1977b) 'Afterword - 1977', *in* Suppe (ed) 1977: 615-730

(1989) *The Semantic Conception of Theories and Scientific Realism* (Urbana: University of Illinois Press)

Suppe, Frederick (ed) (1977) *The Structure of Scientific Theories*, second edition (Urbana: University of Illinois Press)

Tarski, Alfred (1964) 'The Semantic Conception of Truth and the Foundations of Semantics' (1944), *in* L Linsky (ed) *Semantics and the Philosophy of Language* (1952) (Urbana: University of Illinois Press), pp. 13-49

Ullmann-Margalit, Edna (ed) (1988) *Science in Reflection*, The Israel Colloquium: Studies in History, Philosophy, and Sociology of Science, Vol. 3 (Dordrecht: Kluwer)

van Fraassen, Bas C. (1980) *The Scientific Image* (Oxford: Clarendon Press)

Walker, Ralph C.S. (1989) *The Coherence Theory of Truth. Realism, Anti-realism, Idealism* (London / New York: Routledge)

Whorf, Benjamin L. (1964) *Language, Thought, and Reality. Selected Writings of B.L. Whorf* (1956), J.B. Carroll (ed) (Cambridge (Mass): The MIT Press)

Wittgenstein, Ludwig (1992) *Philosophical Investigations* (1953), trans. G E M. Anscombe, reprint of the third edition (1967) (Oxford: Blackwell)

(1994) *Tractatus Logico-Philosophicus* (*Logisch-Philosophische Abhandlung* (1921)). trans. D.F. Pears and B.F. McGuinness, reprint of the first paperback edition (1974) (London: Routledge)

Wolff, Robert P. (1963) *Kant's Theory of Mental Activity. A Commentary on the Transcendental Analytic of the 'Critique of Pure Reason'* (Cambridge (Mass): Harvard University Press)

Wolters, Gereon (1989) 'Phenomenalism, Relativity and Atoms: Rehabilitating Ernst Mach's Philosophy of Science' (1987), *in* J.E. Fenstad, I.T. Frolov, R. Hilpinen (eds), *Logic, Methodology and Philosophy of Science VIII: Proceedings of the Eighth International Congress of Logic, Methodology and Philosophy of Science, Moscow, 1987* (Amsterdam: North-Holland), pp. 641-60

(1990) 'Il positivismo visionario: la filosofia della scienza di Mach', "Intersezioni", 1990, **10**, n. 2, pp. 319-43

INDEX OF NAMES AND SUBJECTS

The University of Western Ontario Series in Philosophy of Science

1. J. Leach, R. Butts and G. Pearce (eds.): *Science, Decision and Value.* 1973
ISBN 90-277-0239-X; Pb 90-277-0327-2
2. C.A. Hooker (ed.): *Contemporary Research in the Foundations and Philosophy of Quantum Theory.* 1973 ISBN 90-277-0271-3; Pb 90-277-0338-8
3. J. Bub: *The Interpretation of Quantum Mechanics.* 1974
ISBN 90-277-0465-1; Pb 90-277-0466-X
4. D. Hockney, W. Harper and B. Freed (eds.): *Contemporary Research in Philosophical Logic and Linguistic Semantics.* 1975 ISBN 90-277-0511-9; Pb 90-277-0512-7
5. C.A. Hooker (ed.): *The Logico-algebraic Approach to Quantum Mechanics.*
Vol. I: Historical Evolution. 1975 ISBN 90-277-0567-4; Pb 90-277-0613-1
Vol. II: Contemporary Consolidation. 1979 ISBN 90-277-0707-3; Pb 90-277-0709-X
6. W.L. Harper and C.A. Hooker (eds.): *Foundations of Probability Theory, Statistical Inference, and Statistical Theories of Science.*
Vol. I: Foundations and Philosophy of Epistemic Applications of Probability Theory. 1976
ISBN 90-277-0616-6; Pb 90-277-0617-4
Vol. II: Foundations and Philosophy of Statistical Inference. 1976
ISBN 90-277-0618-2; Pb 90-277-0619-0
Vol. III: Foundations and Philosophy of Statistical Theories in the Physical Sciences. 1976
ISBN 90-277-0620-4; Pb 90-277-0621-2
7. C.A. Hooker (ed.): *Physical Theory as Logico-operational Structure.* 1979
ISBN 90-277-0711-1
8. J.M. Nicholas (ed.): *Images, Perception, and Knowledge.* 1977 ISBN 90-277-0782-0
9. R.E. Butts and J. Hintikka (eds.): *Logic, Foundations of Mathematics, and Computability Theory.* Part One: Logic, Foundations of Mathematics, and Computability Theory. 1977
ISBN 90-277-0708-1
10. R.E. Butts and J. Hintikka (eds.): *Logic, Foundations of Mathematics, and Computating Theory.* Part Two: Foundational Problems in the Special Sciences. 1977
ISBN 90-277-0710-3
11. R.E. Butts and J. Hintikka (eds.): *Logic, Foundations of Mathematics, and Computability Theory.* Part Three: Basic Problems in Methodology and Linguistics. 1977
ISBN 90-277-0829-0
12. R.E. Butts and J. Hintikka (eds.): *Logic, Foundations of Mathematics, and Computability Theory.* Part Four: Historical and Philosophical Dimensions of Logic, Methodology and Philosophy of Science. 1977 ISBN 90-277-0831-2
Set (9-12) ISBN 90-277-0706-5
13. C.A. Hooker J.J. Leach and E.F. McClennen (eds.): *Foundations and Applications of Decision Theory.*
Vol. I: Theoretical Foundations. 1978 ISBN 90-277-0842-8
Vol. II: Epistemic and Social Applications. 1978 ISBN 90-277-0844-4
14. R.E. Butts and J.C. Pitt (eds.): *New Perspectives on Galileo.* 1978
ISBN 90-277-0859-2; Pb 90-277-0891-6
15. W.L. Harper, R. Stalnaker and G. Pearce (eds.): *Ifs. Conditionals, Belief, Decision, Chance, and Time.* 1981 ISBN 90-277-1184-4; Pb 90-277-1220-4
16. J.C. Pitt (ed.): *Philosophy in Economics.* 1981 ISBN 90-277-1210-7; Pb 90-277-1242-5
17. M. Ruse: *Is Science Sexist?* And Other Problems in the Biomedical Sciences. 1981
ISBN 90-277-1249-2; Pb 90-277-1250-6

The University of Western Ontario Series in Philosophy of Science

18. N. Rescher: *Leibniz's Metaphysics of Nature.* A Group of Essays. 1981
ISBN 90-277-1252-2; Pb 277-1253-0
19. L. Laudan: *Science and Hypothesis.* Historical Essays on Scientific Methodology. 1981
ISBN 90-277-1315-4
20. W.R. Shea (ed.): *Nature Mathematized.* Historical and Philosophical Case Studies in Classical Modern Natural Philosophy. 1983 ISBN 90-277-1402-9
21. M. Ruse (ed.): *Nature Animated.* Historical and Philosophical Case Studies in Greek Medicine, 19th-Century and Recent Biology, Psychiatry, and Psychoanalysis. 1983
ISBN 90-277-1403-7
22. W.R. Shea (ed.): *Otto Hahn and the Rise of Nuclear Physics.* 1983 ISBN 90-277-1584-X
23. H.F. Cohen: *Quantifying Music.* The Science of Music at the First Stage of the Scientific Revolution (1580-1650). 1984 ISBN 90-277-1637-4
24. R.E. Butts: *Kant and the Double Government Methodology.* Supersensibility and Method in Kant's Philosophy of Science. 1984 ISBN 90-277-1760-5; Pb 90-277-2384-2
25. J.R. Brown (ed.): *Scientific Rationality.* The Sociological Turn. 1984
ISBN 90-277-1812-1
26. F. Wilson: *Explanation, Causation and Deduction.* 1985 ISBN 90-277-1856-3
27. J.C. Pitt (ed.): *Change and Progress in Modern Science.* 1985 ISBN 90-277-1898-9
28. H.B. Hollinger and M.J. Zenzen: *The Nature of Irreversibility.* A Study of Its Dynamics and Physical Origins. 1985 ISBN 90-277-2080-0
29. K. Okruhlik and J.R. Brown (eds.): *The Natural Philosophy of Leibniz.* 1985
ISBN 90-277-2145-9
30. G. Oddie: *Likeness to Truth.* 1986 ISBN 90-277-2238-2
31. F. Wilson: *Laws and Other Worlds.* A Human Account of Laws and Counterfactuals. 1986
ISBN 277-2232-3
32. J. Earman: *A Primer on Determinism.* 1986 ISBN 90-277-2240-4; Pb 90-277-2241-2
33. R.E. Butts (ed.): *Kant's Philosophy of Physical Science.* Metaphysische Anfangsgründe der Naturwissenschaft (1786-1986). 1986 ISBN 90-277-2309-5; Pb 90-277-2310-9
34. I.B. MacNeill and G.J. Umphrey (eds.): *Advances in the Statistical Sciences.* Festschrift in Honor of Professor V.M. Joshi's 70th Birthday (6 volumes).
Vol. I: Applied Probability, Stochastic Processes, and Sampling Theory. 1987
ISBN 90-277-2393-1
35. I.B. MacNeill and G.J. Umphrey (eds.): *Advances in the Statistical Sciences.*
Vol. II: Foundations of Statistical Inference. 1987 ISBN 90-277-2394-X
36. I.B. MacNeill and G.J. Umphrey (eds.): *Advances in the Statistical Sciences.*
Vol. III: Time Series and Econometric Modelling. 1987 ISBN 90-277-2395-8
37. I.B. MacNeill and G.J. Umphrey (eds.): *Advances in the Statistical Sciences.*
Vol. IV: Stochastic Hydrology. 1987 ISBN 90-277-2396-6
38. I.B. MacNeill and G.J. Umphrey (eds.): *Advances in the Statistical Sciences.*
Vol. V: Biostatistics. 1987 ISBN 90-277-2397-4
39. I.B. MacNeill and G.J. Umphrey (eds.): *Advances in the Statistical Sciences.*
Vol. VI: Actuarial Science. 1987 ISBN 90-277-2398-2
Set ISBN (Vols 34-39) 90-277-2399-0
40. N. Rescher: *Scientific Realism.* A Critical Reappraisal. 1987
ISBN 90-277-2442-3; Pb 90-277-2528-4
41. B. Skyrms and W.L. Harper (eds.): *Causation, Chance, and Credence.* 1988
ISBN 90-277-2633-7

The University of Western Ontario Series in Philosophy of Science

42. W.L. Harper and B. Skyrms (eds.): *Causation in Decision, Belief Change and Statistics.* 1988 ISBN 90-277-2634-5
43. R.S. Woolhouse (ed.): *Metaphysics and Philosophy of Science in the 17th and 18th Centuries.* Essays in Honor of Gerd Buchdahl. 1988 ISBN 90-277-2743-0
44. R.E. Butts and J.R. Brown (eds.): *Constructivism and Science.* Essays in Recent German Philosophy. 1989 ISBN 0-7923-0251-6
45. A.D. Irvine (ed.): *Physicalism in Mathematics.* 1989 ISBN 0-7923-0513-2
46. J. van Cleve and R.E. Frederick (eds.): *The Philosophy of Right and Left.* Incongruent Counterparts and the Nature of Space. 1991 ISBN 0-7923-0844-1
47. F. Wilson: *Empiricism and Darwin's Science.* 1991 ISBN 0-7923-1019-5
48. G.G. Brittan, Jr. (ed.): *Causality, Method and Modality.* Essays in Honor of Jules Vuillemin. 1991 ISBN 0-7923-1045-4
49. W. Spohn, B.C. van Fraassen and B. Skyrms (eds.): *Existence and Explanation.* Essays in Honor of Karel Lambert. 1991 ISBN 0-7923-1252-X
50. J.C. Pitt: *Galileo, Human Knowledge, and the Book of Nature.* Method Replaces Metaphysics. 1992 ISBN 0-7923-1510-3
51. V. Coelho (ed.): *Music and Science in the Age of Galileo.* 1992 ISBN 0-7923-2028-X
52. P. Janich: *Euclid's Heritage: Is Space Three-Dimensional?* 1992 ISBN 0-7923-2025-5
53. M. Carrier: *The Completeness of Scientific Theories.* On the Derivation of Empirical Indicators within a Theoretical Framework: The Case of Physical Geometry. 1994 ISBN 0-7923-2475-7
54. P. Parrini (ed.): *Kant and Contemporary Epistemology.* 1994 ISBN 0-7923-2681-4
55. J. Leplin (ed.): *The Creation of Ideas in Physics.* Studies for a Methodology of Theory Construction. 1995 ISBN 0-7923-3461-2
56. J.E. McGuire: *Tradition and Innovation.* Newton's Metaphysics of Nature. 1995 ISBN 0-7923-3617-8
57. R. Clifton (ed.): *Perspectives on Quantum Reality.* Non-Relativistic, Relativistic, and Field-Theoretic. 1996 ISBN 0-7923-3812-X
58. P.H. Theerman and K. Hunger Parshall (eds.): *Experiencing Nature.* Proceedings of a Conference in Honor of Allen G. Debus. 1997 ISBN 0-7923-4477-4
59. P. Parrini: *Knowledge and Reality.* An Essay in Positive Philosophy. 1998 ISBN 0-7923-4939-3

KLUWER ACADEMIC PUBLISHERS – DORDRECHT / BOSTON / LONDON

MIX
Papier aus verantwortungsvollen Quellen
Paper from responsible sources
FSC® C105338

If you have any concerns about our products,
you can contact us on
ProductSafety@springernature.com

In case Publisher is established outside the EU,
the EU authorized representative is:
Springer Nature Customer Service Center GmbH
Europaplatz 3, 69115 Heidelberg, Germany

Printed by Libri Plureos GmbH
in Hamburg, Germany